TOWARDS ENABLING GEOGRAPHIES

Geographies of Health

Series Editors
Allison Williams, Associate Professor, School of Geography and Earth
Sciences, McMaster University, Canada
Susan Elliott, Dean of the Faculty of Social Sciences,
McMaster University, Canada

There is growing interest in the geographies of health and a continued interest in what has more traditionally been labeled medical geography. The traditional focus of 'medical geography' on areas such as disease ecology, health service provision and disease mapping (all of which continue to reflect a mainly quantitative approach to inquiry) has evolved to a focus on a broader, theoretically informed epistemology of health geographies in an expanded international reach. As a result, we now find this subdiscipline characterized by a strongly theoretically-informed research agenda, embracing a range of methods (quantitative; qualitative and the integration of the two) of inquiry concerned with questions of: risk; representation and meaning; inequality and power; culture and difference, among others. Health mapping and modeling, has simultaneously been strengthened by the technical advances made in multilevel modeling, advanced spatial analytic methods and GIS, while further engaging in questions related to health inequalities, population health and environmental degradation.

This series publishes superior quality research monographs and edited collections representing contemporary applications in the field; this encompasses original research as well as advances in methods, techniques and theories. The *Geographies of Health* series will capture the interest of a broad body of scholars, within the social sciences, the health sciences and beyond.

Also in the series

Primary Health Care: People, Practice, Place
Edited by Valorie A. Crooks and Gavin J. Andrews
ISBN 978 0 7546 7247 0

There's No Place Like Home: Place and Care in an Aging Society
Christine Milligan
ISBN 978 0 7546 7423 8

Geographies of Obesity: Environmental Understandings of the Obesity Epidemic
Edited by Jamie Pearce and Karen Witten
ISBN 978 0 7546 7619 5

Towards Enabling Geographies
'Disabled' Bodies and Minds in Society and Space

Edited by

VERA CHOUINARD
McMaster University, Canada

EDWARD HALL
University of Dundee, UK

ROBERT WILTON
McMaster University, Canada

LONDON AND NEW YORK

First published 2010 by Ashgate Publishing

Published 2016 by Routledge
2 Park Square, Milton Park, Abingdon, Oxfordshire OX14 4RN
711 Third Avenue, New York, NY 10017, USA

First issued in paperback 2016

Routledge is an imprint of the Taylor & Francis Group, an informa business

British Library Cataloguing in Publication Data
Towards enabling geographies : 'disabled' bodies and minds
  in society and space. -- (Ashgate's geographies of health
  series)
  1. Sociology of disability. 2. People with disabilities--
  Psychology.
  I. Series II. Chouinard, Vera. III. Hall, Edward (Edward
  C.) IV. Wilton, Robert, Dr.
  362.4-dc22

Library of Congress Cataloging-in-Publication Data
Towards enabling geographies : "disabled" bodies and minds in society and space /
[edited] by Vera Chouinard, Edward. Hall, and Robert Wilton.
    p. cm. -- (Ashgate's geographies of health series)
  Includes bibliographical references and index.
  ISBN 978-0-7546-7561-7 (hardback) -- ISBN 978-1-4094-0319-7 (ebook)
  1. People with disabilities. 2. Disabilities. I. Chouinard, Vera. II. Hall, Edward III.
Wilton, Robert, Dr.

  HV1568.T67 2009
  305.9'08--dc22

                                                                    2009045519

  ISBN 13: 978-1-138-24867-0 (pbk)
  ISBN 13: 978-0-7546-7561-7 (hbk)

Contents

List of Figures and Tables

Figures

Tables

Notes on Contributors

Ruth E.S. Allen is currently completing a PhD at the University of Auckland School of Population Health, exploring the meanings of support for childless older people, using qualitative research methods. An experienced clinical psychologist, she is focusing on a developing research career in gerontology, particularly in terms of less pejorative, more nuanced views of the journey of ageing.

Vera Chouinard is Professor of Geography at McMaster University, Canada. A feminist and social geographer, her interests include gendered experiences of oppression and exclusion, cultural representations of those who embody mind and body differences in media such as film, the marginalisation of disabled people in society and space including the impacts of state policy and restructuring on disabled people's lives. Her most recent research focuses on disabled women's and men's lives in the developing nation of Guyana.

Valorie Crooks is an Assistant Professor of Geography at Simon Fraser University, Canada. She has a number of research interests, the most established being geographies of chronic illness and disability. This work has examined women's socio-spatial adjustments to life with chronic illness, including negotiations of paid work and health services. She is presently leading a federally-funded study examining how academic knowledge workers experience the workplace environment when living/coping with multiple sclerosis.

Joyce Davidson is Associate Professor of Geography and Women's Studies, Queen's University, Canada. Her doctoral research on experiences of agoraphobia formed the basis of *Phobic Geographies: The Phenomenology and Spatiality of Identity* (Ashgate, 2003). Her current research and teaching focuses on health, embodiment and different or 'disordered' emotions. She is founding editor of the journal, *Emotion, Space and Society* and has co-edited *Emotional Geographies* (with Liz Bondi and Mick Smith), and *Emotion, Place and Culture* (with Mick Smith, Laura Cameron and Liz Bondi).

Isabel Dyck is Professor of Geography, Queen Mary, University of London. Her research interests focus on issues of gender, body, identity and health. Topic areas include: the transformation of health knowledges and food practices in the context of international migration and the re-making of home; the home as a site of long-term care; and the constructions and experiences of chronic illness and disability.

Books include: Dyck, I., Lewis, N.D. and McLafferty, S. (eds) (2001) *Geographies of Women's Health*, Routledge, London.

Claire Edwards is Lecturer in Social Policy in the Department of Applied Social Studies, University College Cork, Ireland. Her research interests focus on disabled people's involvement in urban regeneration processes, the disability rights agenda, and the relationship between research and policy. She has published on these areas in journals including *Urban Studies, Environment and Planning A, Journal of Social Policy* and *Sociology*, and is currently engaged in research exploring the role of patients' organisations in the governance of medical knowledge.

Edward Hall is Lecturer in Human Geography, University of Dundee, Scotland, UK. One strand of his research focuses on social geographies of exclusion and inclusion of disabled people, particularly people with learning disabilities, and the impacts of changes in welfare state provision. A second strand focuses on the transformation of medical and lay knowledges of health in the context of genetic claims and technologies, including the introduction of genetic healthcare in primary care, and biomedical scientists' conceptualisations of gene-environment relationships.

Louise Holt is Lecturer in Human Geography at the University of Reading, UK. Her research focuses upon embodied inequalities and social reproduction/ transformation, with a specific focus on childhood and (dis)ability. Louise is interested in exploring a range of disabling societal practices that entwine with various mind-body-emotional differences, including diagnoses of disability, special educational needs and 'emotional and behavioural differences'. Louise is editor of *Geographies of Children and Youth, International Perspectives* (forthcoming, Routledge).

Rob Imrie is Professor of Geography at King's College London. He is co-author of *Inclusive Design* (2001, Spon Press, London), author of *Disability and the City* (1996, Sage, London), *Accessible Housing: Disability, Design, and Quality* (2006, Routledge, London), and co-editor of *British Urban Policy* (1993, Paul Chapman Publishing, London; 1999, second edition, Sage, London), *Urban Renaissance* (2003, Policy Press, Bristol) and *Regenerating London* (2009, Routledge, London). He was previously Professor of Geography at Royal Holloway University of London.

Robert J. Kruse is Associate Professor of Geography, West Liberty University, West Virginia. His research interests are in the areas of disabilities, identity and popular culture. His work has appeared in *Area, Canadian Geographer, Journal of Cultural Geography, Journal of Geography, Social and Cultural Geography* and *Southeastern Geographer*. He is the author of, *How Dwarfs Experience the World Around Them: The Personal Geographies of 'Disabled' People* (Edwin Mellen Press, 2007) and *A Cultural Geography of the Beatles* (Edwin Mellen Press, 2005).

Robyn Longhurst is Professor of Geography, University of Waikato, and an editor of *Gender, Place and Culture: A Journal of Feminist Geography*. She teaches feminist, social, and cultural geography. Robyn has published on issues relating to pregnancy, mothering, 'fat' bodies, food, 'visceral geographies', masculinities, and the politics of knowledge. She is author of *Bodies: Exploring Fluid Boundaries* (2008) and co-author of *Pleasure Zones: Bodies, Cities, Spaces* (2001) and *Space, Place, and Sex: Geographies of Sexualities*.

Jo Frances Maddern is a Learning and Teaching Development Coordinator at the Centre for the Development of Staff and Academic Practice and School of Education and Lifelong Learning, Aberystwyth University, Wales. She has an interest in pedagogy in higher education, particularly theories of learning, reflective practice and academic working lives. She has a PhD in geography and also writes on issues of tourism, mobility, bio-politics, heritage, memory and identity.

Deborah S. Metzel is a Lecturer at the University of Massachusetts Boston in the Department of Environmental, Earth and Ocean Sciences. She teaches several broad courses in social geography, where she introduces the social construction of disability. Her research has focused on geographies of employment of people with disabilities. Her research continues to examine geographies of people with intellectual and developmental disabilities (IDD) as she follows the lives, places, and times of the people with IDD of the Baby-Boomer generation.

Hester Parr is Reader in social geography at Glasgow University. She has worked on questions of mental health for many years and has recently published a monograph *Mental Health and Social Space: Geographies of Inclusion?* (Blackwell, 2008). In general terms her work seeks to rescript the social prescriptions surrounding mental health and illness, in partnership and collaboration with varied voluntary sector support services.

Andrew Power is a researcher at the Centre for Disability Law and Policy, University of Galway, Ireland. His current research focuses on the modernisation of the Disability Support Delivery in Ireland. He has ongoing interests in health and social geographies, particularly disability, caregiving, welfare and voluntarism. He is particularly interested in how different national models of welfare 'target' caregivers and people with disabilities. His PhD research examined geographies of family care and how institutional structures at national, regional and local levels weave together to provide care support.

Tracey Skelton is Professor of Critical Geographies at Loughborough University and currently seconded to the National University of Singapore. She has published work on children's and young people's geographies (most recently in *Area* and *Geography Compass*), young D/deaf, lesbian and gay identities, Caribbean studies and research methodologies. She is currently working on a research project at

NUS called 'Youth, Citizenship and Global Futures' which compares young people in Auckland and Singapore. She is the editor of Viewpoints for *Children's Geographies*.

Emma Stewart is Lecturer in Human Geography, University of Strathclyde. Her research interests lie at the interface of forced and skilled migration research, with previous work exploring the asylum-migration nexus in the context of health professional migration. Her current research project is exploring inter-diaspora networks of the Eritrean diaspora and transnational citizenship. Her interest in biometric geographies primarily stems from her ongoing research on refugees and asylum seekers in the UK and the impact of asylum policy upon vulnerable individuals.

Gill Valentine is Professor of Geography at the University of Leeds where she teaches social geography. Gill's research interests include: social identities, citizenship and belonging; and geographies of intimacy. She has co-edited/authored 14 books and over 100 articles. Her research has been recognised through the award of a Philip Leverhulme prize, and the Royal Geographical Society Gill Memorial award.

Janine Wiles is a health geographer and Senior Lecturer at The University of Auckland, New Zealand, in the School of Population Health. Her current research focuses on older people and their environments, and she leads a project on Resilient Ageing in Place. She teaches on topics such as population health, community development, and qualitative research. She previously worked at McGill University in Canada and the University of St Andrews in Scotland, in the fields of urban, social and health geography.

Robert Wilton is Associate Professor of Geography at McMaster University. His teaching and research broadly focuses on social geographies of exclusion, with particular interest in the experiences of people with disabilities. His most recent research examines disabled people's experiences in paid employment. Among his recent publications is: Wilton, R. 2008. Workers with disabilities and the challenge of emotional labour. *Disability and Society* 23(4), 361-373.

Preface

This edited collection represents a 'second wave' of geographical studies of disability. In the decade since Ruth Butler and Hester Parr's *Mind and Body Spaces* (1999), geography's interest in disability has broadened and deepened: other bodies of difference have become included in the disability frame, the bodily experiences of people with impairments have been examined in more detail, the increasing role of technology in disabled people's lives has been studied, and there has been a growing desire to engage with policy debates that affect disabled people's lives. This collection attempts to capture these developments.

Not long before completion, the title of the book was changed to reflect an underlying current present in many of the chapters. Whilst geographers continue to play a key role in examining and critiquing exclusionary spaces and practices, there is a notable and welcome shift towards studying and promoting the means of creating more inclusionary and enabling social spaces for disabled women, men and children. We hope this shift will contribute to the creation of more enabling and empowering lives for disabled people.

Sincere thanks goes to all of the contributors for being part of this project. All worked very hard to complete their work on time, and responded to our comments and suggestions with warmth and enthusiasm. As editors it has been extremely satisfying and rewarding to work with such a talented and committed group of scholars.

Valerie Rose at Ashgate has guided us expertly through the long process from proposal to completed manuscript, patiently responding to our many questions. We also thank our respective institutions and, in particular, our colleagues for support and encouragement in undertaking this project.

Vera thanks her partner Ed, and children Renee and Alex for all their love, support and encouragement. She also thanks Beau Byron. Edward would like to thank his wife Janine, and children Ishbel and Rosina, for their love and support. Robert thanks Cynthia and Sam for their love, support and a steady supply of laughter.

Vera Chouinard, Edward Hall and Robert Wilton, 2010

Acknowledgements

Chapter 2 first published in *Housing Studies*, Volume 19, Issue 5, 2004, 745-763 (reprinted with permission).

Chapter 5 first published in *Information, Communication and Society*, Volume 12, Issue 1, 2009, 44-65 (reprinted with permission).

Chapter 10 first published in *Social and Cultural Geography*, Volume 3, Issue 2, 2002, 175-191 (reprinted with permission).

Figure 7.1 used with permission of Deborah and Charlotte Metzel.

Figure 11.1 used with permission of Pauline (study participant; name changed).

To all those struggling for a more enabling society

Chapter 1

Introduction: Towards Enabling Geographies

Vera Chouinard, Edward Hall and Robert Wilton

Introduction

This collection captures a vibrant and dynamic 'second wave' of geographical studies of disability, which builds on and expands an established sub-disciplinary field of inquiry. It is a field of inquiry, however, with a relatively short history. Despite the prevalence of disability, and the spatial expression of social exclusion and discrimination experienced by many disabled people, geography's interest was marginal until the mid to late 1990s, a 'first wave' of research gathered in Butler and Parr's (1999) edited collection. In this introductory chapter, we trace the development of the geographical study of disability, identifying the distinctive contributions of a spatial interpretation. We then discuss four themes that characterize the second wave of disability research in geography. The chapter ends with consideration of possible directions for future geographic research on disability.

There have been several comprehensive reviews of geographic research on disability, which we have no intention of repeating here (e.g., Chouinard 1997, Park, Radford and Vickers 1998, Imrie and Edwards 2007). However, it is important to reflect briefly on the development and contribution of a spatial interpretation of disability. Prior to the first wave of critical research, geography's interest fell into three camps. First, the positivistic tools of epidemiology, adopted by medical geographers, were applied to the incidence and distribution of disabling conditions, such as multiple sclerosis (Mayer 1981) and spina bifida (Lovett and Gatrell 1988), with the intention of understanding causal factors. Here, 'disability' is an individual diseased body, and 'geography', a non-defined background within which bodies are located.

A second set of studies began to examine these contexts, concerned with access and mobility for disabled people within, in particular, urban environments, focused on barriers presented by poorly designed transport systems and urban landscapes (Hahn 1986, Gant and Smith 1988). Golledge approached this in a different way, adopting a behavioural interpretation to examine how people with visual impairments understand (what he termed 'spatial cognition') and engage with urban environments, and how technical devices such as GPS and GIS can be applied to allow people to move around the hazards of a city (Golledge et al. 1991, Golledge 1993). These studies did not examine the underlying structural and institutional production of the barriers people encounter, yet their descriptions revealed in great detail the extensive spatial exclusion that many disabled people experienced in city spaces.

The third distinct set of studies was concerned with mental health, again adopting medical geography's approach with dual interests in the spatial distribution of mental health problems and the location of mental healthcare services. For example, Giggs (1973, 1986) attempted to explain the incidence of schizophrenia in Nottingham, UK, making connections between mental health problems and poor inner city living conditions, while other work (e.g., Dean and James 1981) began to explore the broader links between mental health and socio-spatial contexts. The deinstitutionalization of people with mental health problems out of large scale hospitals and asylums into community environments stimulated a large number of studies concerned with the location of and (lack of) support for people in urban communities, the rise of homelessness and poverty among former 'mental patients', and the response of mainstream communities to the housing and facilities for people with mental health problems (e.g., Dear and Taylor 1982, Dear and Wolch 1987). In these studies there is a powerful capturing of the sense of 'difference' experienced as people with mental health problems entered spaces that they had played no part in shaping, and the consequential discrimination and exclusion.

In summary, geography was active in its examination of disability prior to the so-called first wave of studies. Whilst a stinging critique of this early work was soon to emerge – centred on an individualistic conception of disability and a lack of concern with broader socially and spatially produced causes of exclusion – it is important to emphasize that clear spatial insights were achieved, most notably into the prevalence of disabling conditions, the significance of social and cultural environments in shaping the lives of disabled people, and the embedded discrimination in mainstream communities and in welfare services. Such insights were highly significant for the subsequent 'critical' geographies of disability.

The juncture at which these new critical or interpretative studies arguably began to emerge was an exchange in the pages of *Transactions of the Institute of British Geographers* in the mid-1990s, centred on the conceptualization of 'disability' and the ways in which geographers should approach the issue. Influenced by broader societal changes – in particular the increasing political activism of some disabled people and the associated theoretical developments inspired by disabled academics – geographers began to critique earlier geographical studies of disability and lay out an agenda for research. Butler (1994), Gleeson (1996) and Imrie (1996a) were sharp in their critique of Golledge (1993), claiming that he – and by extension the majority of the research done in the preceding years – failed to address the inherently 'disabling' nature of the socio-spatial environment, focusing instead on the impairments of the individual and consequent difficulties faced in these 'distorted' and 'transformed' spaces (Golledge 1993: 64). In addition, the critique of Golledge focused on the 'ableist' nature of the research, noting the lack of presence of the voices and experiences of disabled people in the data, advocating instead a geography *with* rather than *for* disabled people (Gleeson 1996). Rethinking disability as a socially and spatially produced form of exclusion and oppression, and recognizing the capacity of disabled people to challenge or transform such disabling social relations and spatial structures, formed the conceptual core of the first wave of disability studies in geography (see *Environment and Planning D* 1997, Butler and Parr 1999).

Developing the concern with barriers to mobility in urban locations, new critical geographical studies of disability shifted the focus from individual 'inability' to navigate, to the socio-political construction of disability within, and through, the built environment (Laws 1994, Imrie 1996b). Gleeson (1999a), from a historical materialist perspective, stressed the manner in which disability was produced by 'disabling' environments – expressed by Butler (1994: 368) in the phrase 'the blind live in a world built by and for the sighted' – and economic and care systems. Butler and Bowlby (1997), arguably revisiting the concerns of Golledge in their work on people with visual impairments in public space, connected a critique of environments imbued with disablist attitudes with the experiences and embodied feelings of people of being in such environments (and indeed their reactions, responses and resistance). The focus on the disabled body, on the experiences of impairment within particular spaces, was a central theme of the first wave (Parr and Butler 1999). Drawing on postmodern interests in the body and individual subjectivity and identity, many studies refocused the spatial concern to microgeographies, including Hall (1999) who considered embodied experiences and identities of disabled people in employment, and Dyck (1995) on the 'lifeworlds' of women with multiple sclerosis, and their renegotiation of their identities within the spaces of the home. These studies reflected a concern with the overly social constructionist nature of the social model, which risked denying the material, embodied realities of impairment (Hall 2000). In this sense, the continued assertion of the material, of the lived experiences of impaired bodies in particular places, has also been a major theme and contribution of a geographical perspective.

A final theme which defined the first wave of studies was an increasing concern with the relations of research, in particular the 'ableism' of the discipline itself which through its research practices and broader social relations 'constructed disabled people as marginalized, oppressed, and largely invisible, others' (Chouinard 1997: 380). Such assumptions and conceptualizations combine with, and are a reflection of, research that treats disabled people as objects for research.

Geography has broadened and deepened its interest in disability in the second wave of research, represented in this collection. Whilst research has continued on the still significant material exclusion and discrimination experienced by disabled people in public and private spaces (e.g., Wilton 2004, Driedger, Crooks and Bennett 2004), the appreciation of the relationship between disability and space has become more nuanced and complex. The following sections set out what we consider to be the central themes of the second wave of research, drawing on the chapters in the collection: first, a broadening of the meaning of 'disability' to encompass other bodies and experiences; second, a deepening concern with the embodied experiences of disability and chronic illness; third, the possibilities and challenges of the increasing interaction between disabled people and technology; and fourth, an interest in shaping policy agendas and the 'place' of disabled people in contemporary society.

The Meaning of Disability

A shift of focus in the second wave to the relationship between embodied selves and a range of spaces has produced an almost unnoticed widening of the concept of 'disability' in geography (Crooks, Dorn and Wilton 2008). The insights of a spatially-infused interpretation of disability, as this collection shows, are increasingly being applied to other bodies of difference, including older people (Milligan 2004, Wiles and Allen Chapter 12), 'fat' people (Colls 2006, Longhurst Chapter 11), people of small stature ('little people') (Kruse Chapter 10), people with intellectual impairment and learning difficulties (Hall 2004, Davidson and Parr Chapter 4, Metzel Chapter 7, Power Chapter 6) and emotional problems (Holt 2004, Chapter 8), and people with chronic illness (Del Casino 2001, Crooks 2007, Chapter 3). This development answers Parr and Butler's (1999: 10) call for a broadening of the scope of disability geography to encompass the embodied experiences of 'all sorts of different people, with all sorts of different mind and body characteristics.' However, a broadening of 'disability' is not without difficulties: Parr and Butler recognized that it could dilute the power of the disability concept (and the political movement closely linked to it), especially when some do not claim the disability label; in addition, there is the necessity in particular circumstances to specify impairments and disabilities (for example, in demands for appropriate support); and there remain some, for example, D/deaf people, who assert an identity position distinct to and in tension with the 'disabled' identity (Valentine and Skelton Chapter 5).

Crooks (Chapter 3) examines how the onset of a chronic illness, Fibromyalgia Syndrome, transforms the lives of a group of women, in particular the experiences of home spaces, and their roles and relationships within them. Crooks employs the concept of 'disablement' to interpret the (re)negotiation of spaces of the home and beyond, by focusing on 'the interconnections between the mind, the body and lived realities' (p. 54). For many, the interactions of embodied limitation and social barrier produced what could be termed a 'disabling' position of social and spatial isolation within the home. There are consequences to this approach, however: incorporating experiences of chronic illness into a disability frame necessarily means considering matters of illness and medicine, and fluctuations in bodily impairment, questioning the dominant social model of disability (Crooks, Chouinard and Wilton 2008).

The focus on the *process* of disablement, rather than on a disabled *identity*, is also present in the contribution by Longhurst (Chapter 11). She argues that mainstream spaces are often disabling, in emotional as well as material ways, for women who identify as 'fat', in the confinement and restriction of movement, and the socio-cultural non-acceptability of larger bodies. There are sufficient overlaps of experience – for example, 'both fat people and people with disabilities are inscribed by medical processes [and] can experience social isolation' (p. 202) – for fat people to be seen as experiencing disabling processes and environments that venerate normal bodies (thinness and ableness) and see 'dissident bodies' (p. 213) as abject. The adoption of a disabled identity by fat people (advocated by Cooper 1997) is complex since in most people's eyes disabled people are blameless for their situation, whilst fat people are deemed responsible for theirs. The participants in Longhurst's study were not asked directly if they considered themselves to be disabled, and none claimed such an identity though their experiences were similar (this suggests that the wider use of the 'disablement' process to make sense of other bodies of difference is in most cases not accompanied by the affirmation of a politicized disabled identity). A reinterpretation of bodily difference as disability does provide an opening to reconceptualize the 'problem'; in the case of fatness, geographical studies are at present concerned with individual and group behaviour producing 'obesity'; rethinking this through the notion of disability allows social and cultural environments to be considered as significant factors in producing fatness.

The issues of applying the label of 'disabled' to a group that until now has not been so named are also considered by Wiles and Allen, in their examination of ageing (Chapter 12). They note that whilst a significant proportion of the people who are classified as 'disabled' are older people, the disability label is commonly not attached to them. Instead, their impairments are understood as an inherent part of the ageing process. Wiles and Allen argue for a focus on the processes that occur in socio-spatial contexts that exclude older people physically, socially and culturally; though they concede that this has the troubling side-effect of marking out disability as a negative identity. Importantly, Wiles and Allen argue that insights from disability literature can inform studies of ageing; for example, in the recent embracing of an experiential, embodied model of ageing. The reverse is also true: disability geographies can learn from the way in which, with ageing, bodily impairment can be seen as an integral element of self-identity.

Described as a 'small corpus' of work within the geographies of disability (Wolch and Philo 2000: 138, cf. Hall and Kearns 2001), studies of learning or intellectual disability are well represented in the second wave, with contributions in this collection from Power (Chapter 6) and Metzel (Chapter 7). Both chapters examine the central role that caring relations play in the lives of people with learning disabilities. Power critiques the dependency/independence discourse in favour of the notion of 'interdependence', which better reflects the particular socio-spatial lives of people with learning disabilities. It is the relationships between individuals, family members and carers that generate both barriers to, and possibilities for, living in the community and strengthening self and collective identity. Metzel, starting from personal experience, investigates those who care for people with learning disabilities at a distance. In both of these chapters, the extension of research into the lives of those *without* disabilities recognizes the mutual constitution of the lives of disabled and non-disabled people, and of the disability category itself.

As we suggested above, studies of mental health have a long history in disability geography. A decade ago, Parr and Butler (1999: 12) argued that the evolving field of disability research should embrace a broad range of mental differences, reflecting 'the multitude of embodied and behavioural characteristics which are seen as socially stigmatizing and amenable to medical categorization and treatment'. Authors in the second wave have responded to this call, addressing new themes of mental difference that lie outside of what has generally been considered mental health and impairment. Davidson and Parr (Chapter 4), for example, consider individuals on the Autistic Spectrum and those with anxiety disorders, who commonly experience society as emotionally and sensorally challenging, as well as experiencing social stigmatization. People with these conditions can, however, Davidson and Parr argue, assume a positive collective identity as 'new neurological minorities' (p. 78) through the medium of the Internet.

Increased interest in other mental differences is also evident in Holt's (Chapter 8) study of young people in education with emotional and behavioural difficulties. Holt argues directly that models of disability should be 'reappraised' to include emotional differences with the intention of strengthening the position of young people with these issues (who are commonly understood through the individual, medical model). She understands disability here as a product of dynamic and ongoing interactions – drawing on notions of 'performativity' – between minds/bodies and specific cultural/social configurations. Including children with emotional and behavioural problems within the 'disability' frame thus allows the 'problem', in the same way as obesity/fatness, to be seen as a product of interaction between the individual and the school, not solely the result of an inherent bodily/emotional inadequacy. This also allows those who have not been medically diagnosed to be identified as 'disabled'; a tool of strategic essentialism to gain recognition and support, but which runs counter to dominant understandings of disability.

The examination of 'a fuller range of mind-body-emotional differences' (Holt p. 146) under the banner of disability in the second wave of studies is both enlightening and potentially problematic. The spatial lives of many more people with mind/body differences are now being considered by geographers using the conceptual lens of the disablement process and new insights from geography are now being applied to studies of ageing, fatness, and emotional conditions. This relational take on disability has also encouraged the study of non-disabled people who are central to the disability story, for example, families, carers, teachers, and other social actors who help to shape people's lives.

There remain doubts, however, among many people who identify as disabled about the application of the (hard fought for) 'disability' term to describe the situations and experiences of people with a broadening range of bodily differences. Moreover, the use of the disablement process to make sense of people's lives has often not been accompanied by an embracing of the disability identity by people living with other mind/body differences, suggesting that 'disability' remains a largely stigmatized concept in society.

Bringing the Body Back In: Embodiment, Identity, Disability and Spaces of Everyday Life

Over the past decade, disability studies scholars and geographers have been concerned to 'bring the body back in' to our understanding of experiences of impairment and disability. This has encouraged a shift away from a social model of disability toward more embodied conceptions of the lives of persons contending with impaired or ill bodies and minds and processes of disablement in society and space.

From the Social Model to Embodied Conceptions of Impairment and Disability

By any account, the social model of disability has played an important role in challenging biomedical models of disability. Whereas biomedical models conflate an individual state of bodily dysfunction or impairment with disability, the social model distinguishes between impairment as a bodily state and disability as a social process of oppression and exclusion (e.g., Oliver 1990, Gleeson 1998, Chouinard and Crooks 2005).

While recognizing the conceptual and political importance of the social model, scholars have been increasingly critical of the model's binary distinction between impairment as a pre-social biological state and disability as a social process. Hughes and Paterson (1997: 326) argue for 'a critical interrogation and collapse of the concepts of impairment and disability' with disability seen as embodied and impairment as social. Conceptualizing suffering not as personal tragedy but as something 'on the threshold between pain and oppression' helps to illustrate this approach. For example, experiences of ableist employment practices become embodied in terms of subjective suffering and hurt. Experiences of impairment can be seen as social in terms of meanings ascribed to impaired bodies and minds (e.g., as 'lacking') and in that the social conditions of life, such as poverty, malnutrition and war, cause and exacerbate physical and mental impairment (Shakespeare 2006).

A wide range of scholars has contributed to more embodied conceptions of experiences of impairment and disability. Papadimitrious (2008) explores the embodied process of becoming 'enwheeled' – becoming a skilled wheelchair user after spinal cord injury. She argues that the wheelchair becomes incorporated into a user's corporeal schema becoming an extension of the user's embodied being and way of negotiating society and space. For the wheelchair user, 'enwheelment' contributes to an identity of being 'newly able' or renabled; an identity that is at odds with ableist readings of wheelchair use as 'incapacity' and lack. Similarly, Zitzelsberger (2005) shows how women with physical impairments struggle to resist ableist readings of embodied attributes such as facial disfigurement by, for example, affirming such differences as a 'work of art'.

Geographers have also contributed to efforts to think about experiences of impairment and disability in more embodied ways. Hall (2000: 206) has argued for 'in-between' conceptions of the body and impairment in which the body is seen as social and 'social processes as part of the body, or embodied.' More recently, Hansen and Philo (2007) explore how experiences of impairment are complicated by social normative pressures to cope with the impaired body in as 'normal' a way as possible: for instance to walk without aids such as crutches. Geographers have also explored the embodiment of mental (ill-)health. Parr (2000) has argued for closer attention to how embodied performances of the mentally ill self in place and the meanings assigned to these performances construct social and spatial boundaries between more and less acceptable ways of being 'mad'. Embodied expressions of being mentally ill which are not threatening (e.g., staring into space) are likely to be read as acceptable whilst more threatening performances such as violent behaviour are likely to prompt efforts to socially and spatially exclude. Reflecting the emotional turn in geography, Smith and Davidson (2006) discuss ways in which phobias of natural objects (e.g., spiders) are embodied: experienced through emotions such as disgust and fear, bodily sensations such as a sufferer's 'skin crawling' and apprehension about contamination through the feared object transgressing bodily boundaries.

Embodied Geographies of Impairment and Disability: The Present Collection

The chapters in this collection also contribute to more embodied, geographic understandings of impairment, disability, identity and difference. Davidson and Parr (Chapter 4) consider how those who embody the neurological and communicative differences of autism and of mental health challenges struggle to create more enabling spaces of encounter online. For those who embody autistic differences virtual space offers a more enabling and less stigmatizing form of textual encounter where autistic differences can be celebrated rather than devalued. Virtual encounters can be both enabling and emotionally risky for those with mental illness: creating spaces of support but also necessitating strategies of 'boundary maintenance'. In Chapter 5, Valentine and Skelton explore some of the paradoxical implications of Internet use and virtual communication for persons who identify as D/deaf. They argue that virtual communication has helped to diminish the exclusion of persons who are D/deaf from hearing society; opening up direct access to information not available through traditional, off-line oral communications. Yet Internet use has paradoxical implications: while increasing access to information it does not integrate D/deaf people into the hearing world because hearing people are not required to change their communication practices to accommodate D/deaf people's needs (e.g., through use of virtual sign language).

Dominant, idealized constructions of the meaning of home represent it as a place of sanctuary, independence, privacy and well-being. Such representations, however, neglect embodied aspects of the encounter with spaces in the home such as bodily impairment and the negotiation of physical designs inconsistent with the needs of those whose corporeality is 'other' than able. In Chapter 2, Imrie explores embodied encounters with the home space and the meanings assigned to the home by persons with mobility impairments. For some the home was a place of entrapment when, for example, they were confined to it for extended periods while awaiting modifications to its physical structure. The home was also experienced as a space of dependence when assistance was necessary, for example, to reach items in kitchen cupboards too high to be accessed by a wheelchair user. Nor was the home necessarily a private space if, for example, the necessity of getting a wheelchair into a small bathroom meant that the bathroom door could not close. Imrie stresses that disabled people are not passive victims of unaccommodating domestic spaces; rather they actively modify those spaces of life including strategies of physical redesign or using only the first floor of a home. Imrie makes a convincing case for more embodied conceptions of 'other' ways of negotiating and assigning meaning to the home. In Chapter 3, Crooks also examines the embodied negotiation of home space focusing on the experiences of women with Fibromyalgia. The domestic space of the home loomed larger in these women's lives and identities through the loss of an embodied presence elsewhere, notably in the workplace. She stresses that these women's embodied negotiations of the home were emotionally complex and charged. For example, while the home was experienced as a place of rest and comfort, it was also a place of unease as the women struggled to come to terms with the increased prominence of domestic spaces of the home within their lives and the loss of prior identities and roles in and outside the home. It was also a space of changing social identities and roles as, for example, the women experienced difficulties performing domestic labour in accord with gendered expectations.

Chapters in this collection also help to broaden the range of embodied differences understood to have disabling consequences. Longhurst (Chapter 11) examines the emotional and material geographies of women who are large or 'fat'. For the women she interviewed being larger than slender ideals was an emotion-laden experience: for example, feeling emotionally devalued by others, and facing fears that their embodied presence in places such as airplanes would encroach on other people's personal space and that others would negatively judge their embodied selves. Such feelings contributed to spatial avoidance strategies such as not frequenting beaches. Longhurst shows that environments for these women are also disabling in a material sense when, for example, theatre seats or changing rooms are too small to comfortably accommodate being 'large'.

Kruse, in Chapter 10, also considers issues of body size and disabling environments. Focusing on the experiences of 'little people', he argues that the 'staturization' of space for those of average height helps to marginalize and disable people of short stature in many spaces of everyday life. He points out that little people do resist the staturization of space; for example with sufficient income they may be able to re-scale the private spaces of the home to better accommodate their stature.

Ageing is another basis upon which people come to experience, in embodied ways, being 'different'. In Chapter 12, Wiles and Allen discuss some of the ways in which theoretical debates about ageing parallel those in the disability studies literature. They note, for example, how ambivalence about claiming identities of 'old age' parallels how people relate to 'disabled' identities. In both cases, stigma plays a role in constructing these identities as an undesirable embodied difference. Debates about how to conceptualize ageing also parallel those concerned with disability: ranging from biomedical models to social constructs and embodied difference. Wiles and Allen also note conceptual and political tensions between addressing issues of inequality with regard to experiences of being 'old' and celebrating the difference of 'old age' – something that also resonates with the disability studies literature. Akin to recent developments in disability geography, Wiles and Allen argue that conceptualizing ageing as an embodied process which unfolds in place and over space offers a promising way of drawing these strands of debate together. In such an approach bodily experiences and socio-spatial forces are included in an understanding of how the 'difference' of old age is produced, embodied and experienced.

In the final chapter of the book, Isabel Dyck considers some of the implications of advances in biomedical and technological 'body knowledge' for geographies of impairment and disability. She argues that techniques such as prenatal genetic screening and stem-cell selection have taken the biomedical gaze to the molecular level of embodied life – posing issues such as what lives are to be valued and who is to manage life itself. She suggests that geographers are well placed to contribute to our understanding of these socio-spatial processes of 'body optimisation.' There is a need, for example, to explore issues of access to new techniques and the sites and spaces in which they are being put into practice. This would include consideration of the extent to which the application of this knowledge is pushing the boundaries of what constitutes 'normality' in terms of embodied life and ways of being in the world. As Dyck's chapter indicates, there are important and challenging new horizons for geographic research concerned with impairment, disability and the management and modification of embodied life.

Technology and Disability

Disabled people's relationships with technology are complex and contradictory, not least because technology has the capacity to impair and disable on the one hand, and to reduce impairment and enable participation on the other (Sheldon 2004). These complex relations are reflected in the ways scholars have theorized technology in relation to disability. Some authors have positioned technology as an inherently liberating phenomenon. Finkelstein (1980), an influential voice in disability studies, argued that advances in technology were critical to the liberation and economic independence of impaired people.

However, more critical interpretations have cautioned that there is nothing inherently liberating about technology since it is a profoundly social phenomenon; its design, production and availability is shaped by prevailing political, economic and cultural conditions (Sheldon 2004). As Gleeson (1999b) has argued, misplaced faith in the liberatory potential of technology stems from a failure to theorize the broader cultural and material conditions that oppress disabled people in capitalist society. At the same time, recognizing the socially embedded nature of technology directs our attention to the ways that disabled people struggle to influence the design, availability and uses of technology to foster more enabling environments. Two areas are of particular relevance to this volume: information technology and recent developments in genetics and biotechnology.

Information Technology

Recent work has emphasized the potentially enabling capacity of information technology such as the Internet and information and communication technologies (ICTs). At the same time, there are concerns that the popularity of virtual environments constitutes a 'technical fix' that reduces the need to make meaningful change to the physical and social environments beyond the disabled person's home (Sheldon 2004). Two chapters in this collection are centrally concerned with the extent to which different disabled groups are able to use the Internet as a setting for social interaction, information sharing, identity construction and political advocacy.

Valentine and Skelton examine D/deaf people's use of the Internet and its implications for their access to information and their communication with other D/deaf people and the larger hearing society. Their research effectively demonstrates the complex and sometimes contradictory effects of technology in the lives of D/deaf people. On the one hand, the Internet provided tremendous opportunities to combat the 'information poverty' faced by many D/deaf people as a result of more traditional communication technologies. The availability of information provided opportunities for empowerment in political participation, health care and other social spheres. The Internet also provided many positive opportunities for developing social relationships and building community among D/deaf people. On the other hand, access to information via the Internet is often achieved at the expense of face-to-face contact with hearing people in offline environments. Moreover, a significant number of respondents reported not revealing their D/deafness in online exchanges, effectively 'passing' as hearing.

The study also reminds us of the diversity within and between disabled populations. D/deaf people's experiences with ICT may be distinct from other populations. Indeed, Valentine and Skelton report that a significantly *higher* percentage of D/deaf people reported daily use of the Internet compared to the general population in the UK. This finding stands in contrast to other studies that have suggested disabled populations in general face difficulties bridging the 'digital divide', and is perhaps an indication of the importance of the Internet as a potentially enabling technology for some D/deaf people. At the same time, D/deaf people vary in their relation to ICT. While some people are highly knowledgeable users, others face multiple barriers to Internet access and use, including income constraints and low information literacy.

Similar themes are explored by Davidson and Parr in their chapter on the Internet as a space for 'enabling cultures' among people with mental health problems and people on the Autism Spectrum. Like D/deaf people, evidence on the experiences of both groups suggests that the Internet can have a number of enabling effects, creating culture, facilitating social networking and fostering political advocacy and activism. For people on the Autism Spectrum, the authors note that the Internet has given people a space from which to speak – both individually and collectively – about their experiences and identities. In this sense it offers a powerful medium to challenge stereotypes of Autism as non-communicative and asocial, and to forge a culture of difference. Yet this is clearly not the whole story. Questions arise as to whether the Internet constitutes a space of resistance to prevailing 'neurotypical' norms, or whether it provides the technology to overcome limitations, to sound 'normal'. For people with mental health problems, online environments like discussion forums can provide valuable spaces away from disabling mainstream settings and the promise of social support. Yet encounters with others dealing with mental ill-health can be intense and difficult, fostering 'new kinds of uncomfortable proximities.' Moreover, the strategies people use to manage these proximities may in fact lead to forms of 'othering' within supposedly accepting online communities.

Collectively, these findings support the contention that there is nothing inherently empowering or disempowering about technology; these engagements with information technology produce multiple, sometimes contradictory, effects. Moreover, these effects themselves are dynamic, shifting over time as conditions change, and as people – individually and collectively – embrace, occupy and unsettle virtual environments.

Genetics and Biotechnology

Another area that has generated considerable debate and discussion in recent years has been the field of genetics and the scientific and technological developments linked to the mapping of the human genome. Wilson (2007) examines the discursive construction of the human genome project and genetic medicine more broadly arguing that its very purpose is to 'correct' errors in the genetic 'instruction book' that result in disease and disability. He argues that this seductive representation of genomics reinforces the stigma of disability as something inherently wrong or defective. This conception of disability underlies the expansion of prenatal genetic testing, which raises profound questions about technology's capacity to identify and selectively abort foetuses that exhibit 'defective' genes.

However, some scholars have argued that critical disability research cannot simply oppose or ignore developments in genetic medicine and genomics (Shakespeare 2005a, Dyck Chapter 14). Indeed, debates about how to engage or respond to these developments link with broader conceptual challenges. Chief among these has been the call to rethink the social model of disability through a re-engagement with the 'realities' of the body. But as Isabel Dyck asks in this volume, what kind of body is to be engaged in an era when science and technology are producing new knowledges and new forms of expertise for managing bodies? For Shakespeare (2005a, 2005b), scholars in disability studies have important work to do in distinguishing between appropriate and normalizing medical therapies and exploring the implications of new treatments. This call is echoed by Dyck who suggests that recent biomedical and technical advances challenge the research agenda of disability geography. Drawing from Rose's work on technologies of optmisation, she contends that research on sites and spaces associated with these technologies will allow us to demonstrate 'the situatedness of knowledge and the very material and social conditions that mediate the outcomes of advances in technology and scientific medicine' (p. 259).

Other questions arise here in relation to technology, disability and inequality. In particular, the vast resources devoted to genetic medicine and genomic research have to be understood in relation to the continuing lack of access to basic health care faced by many people in developing countries. These profound socio-spatial inequalities have direct relevance for geographies of disability since the lack of access to basic health care is responsible for illness and impairment that could be easily prevented in other contexts (Hubbard 2007). In this sense, the ethics of genomics and genetic testing are inextricably tied to questions of social justice and inequality at multiple scales. For Dyck, these linkages require researchers to find ways to bridge between the social and the biomedical, and between illness, impairment and disability in ways that challenge key political commitments of the disability rights movement but that also provide potentially important opportunities for conceptual innovation.

Questions of technology and its relation to social justice at multiple scales also connect to the chapter by Maddern and Stewart. These authors examine the use of biometric technologies to regulate human movement across national borders. They are paritcularly concerned with the ways in which these technologies are used to delineate 'biologies of culpability', which in turn facilitate the exclusion of certain bodies from the privileged status of citizenship and the privileged spaces of certain nations. There are interesting parallels with the discussion of genetics above in that both genomics research and the field of biometrics assume a certain 'truth' located in the biology of the body. Moreover, both reproduce the notion of a single 'normal' body. In the case of biometrics, failure to account for human diversity produces a technology that does not recognize, and consequently delegitimates, bodily difference and disability.

The design of this technology also speaks to broader assumptions and judgements about who can travel where legitimately. These assumptions are a reflection of the security concerns of Western nations in the context of the 'war on terror' (Häkli 2007). However, as Ingram (2008) notes, notions of security in this context extend well beyond terrorism to intersect with the spectre of uncontrolled immigration and the 'burden' of asylum-seekers in the West. There are important conceptual and political overlaps here between the experiences of ill and impaired immigrants and other groups such as poor people of colour constructed as threats to Western countries and the 'privileged domestic citizen body' (Ingram 2008: 876). In this context, processes of exclusion are facilitated by technologies capable of detecting bodies deemed too ill or impaired to qualify for admission.

Politics and Policy

The first wave of geographical studies of disability emerged in the heat of the political and conceptual transformation of 'disability'. The conceptualization of disability as individual pathology was challenged by a dynamic disability movement and the voices of a small grouping of disabled academics and commentators, shifting conceptions of the cause of disability to social structures and institutions (Oliver 1990). Embracing this transformation, the first wave was characterized by a critical examination of disabling environments; an appreciation of the diversity of impairment and other differences (e.g., gender, age and sexuality) within the disabled population; and the inclusion of disabled people's voices and experiences that had been marginalized in previous research.

That the voices of disabled people feature heavily in this collection testifies to the progress made in the second wave of studies. Chapters by Imrie, Crooks, Holt, Longhurst, and Davidson and Parr, for example, place the experiences of people with mind and body differences centrally in their analyses, using techniques including participatory observation, focus groups, face-to-face interviews and online interviews to sensitively represent participants' experiences and views. The second wave of scholarship has shown that partnership with disabled people and their organizations is at the heart of research that fully represents the experiences of people with 'disabling' mind and body differences, and provides tools and information for change. It is noteworthy, however, that 'participatory action research', advocated as an empowering strategy, with disabled people shaping and undertaking research (Kitchin 2001), does not seem to have become established in geographies of disability.

In the second wave there have been efforts to combine critiques of disabling environments with analyses of and engagements with policy, for example on social exclusion and inclusion (Hall 2004), housing design (Imrie 2006) and employment (Wilton 2004). Yet, Imrie and Edwards (2007) note geography's absence in policy development relating to disability, a theme pursued by Edwards (Chapter 9). She observes that since the 1997 election of the Labour Government in the UK, there has been an emphasis on 'evidence based' research or 'what works'; for disabled people this has also been a period of intensive policy making, from care provision to promoting employment, making these evaluations important for people's lives. Edwards, echoing Cameron's (2005) concern about geography's irrelevance in shaping policy, uses the example of the 'welfare to work' policy of the Labour Government to demonstrate how current evaluations of policy have been actively despatialized. To reclaim these 'absent geographies of disability' she highlights how individual and local scale experiences of disabling economic environments must be addressed, as well as the local and national outcomes of welfare policies, citing Mohan's call to recognize 'the difference that space makes to the operation of welfare systems' (2003: 370). However, there remains uncertainty within geography about being part of such evaluations, with some advocating instead 'counter-policy research', critiquing policy from the outside (Pain 2006: 251) or engagement with other forms of 'policy' pursued by voluntary and campaigning organizations (Ward 2005).

Chapters in the collection on education and care demonstrate that geographers have begun to engage in such critical yet policy relevant work. Holt (Chapter 8) makes a case for incorporating young people with emotional and behavioural difficulties into the disability frame, to better conceptualize and politicize their experiences as socially produced. She argues that this conceptual shift provides an important opportunity to challenge how current educational models and practices understand and treat children with such behaviours. The chapter also highlights that the notion of 'special educational needs' is poorly defined in institutional guidelines, and spatially variable between schools and areas. Such definitional uncertainty is exacerbated by policies used to manage children, primarily exclusion, segregation and punishment. Holt argues for more inclusive school spaces, perhaps utilizing some segregated spaces, and less hierarchical relationships between teachers and pupils.

The policies of, and approaches to, care are explored by Power, Metzel, and Wiles and Allen (Chapters 6, 7 and 12, respectively). Power critiques the dualism of dependence/independence in the debate around the provision of care for people with intellectual disability; he argues instead for recognition of the complex interdependencies that people form with families, carers, supporters and friends in social networks. The expansion of 'person-centred planning' for decision-making on individual care holds much potential for improving people's lives. Yet Power cautions that this approach has not taken into account the broader structures and barriers that constrain lives, and also the complex social networks of which people with intellectual disability are part. Without due regard for these contexts of people's lives, empowerment cannot be achieved. Metzel also considers family caring relations, reflecting on her own and others' experiences of long distance sibling care for persons with intellectual disability. Caring is usually thought of as an intimate and proximate relationship; Metzel explores how the challenges of distance produce new forms of caring, through telephone calls, letters and paintings, but with an undercurrent of guilt for being physically absent and a sense of intense responsibility for the family member. In an era of deinstitutionalized care and dispersed families, the spatiality of care will continue to be a challenge for the provision of care and those involved in caring relations. Wiles and Allen further the debate on caring relations, identifying the need to examine embodied ageing and disability 'in place'. More older people now remain in their homes and communities as they age, receiving local support to encourage independence, rather than residential care. However, some local places do not welcome and adapt to those who are ageing, granting them little say in how their community develops.

Providing care in private homes, rather than public institutional settings, can also be isolating for those receiving and giving care.

As noted above, the politics of disability identity were central to the first wave of studies. Research in the second wave reflects the broadening and growing complexity of the concept of 'disability', raising questions about identity and citizenship for people with bodies of difference (and the coherence of the disability political movement). Older people, for example, as Wiles and Allen note, occupy an ambivalent position in society, often simultaneously seen (by themselves and others) as physically and mentally impaired *and* as sources of wisdom and experience. Others who have become incorporated into the disability field – 'fat' people (Chapter 11), people with chronic illness (Chapter 3) and young people with emotional and behavioural difficulties (Chapter 8) – likewise occupy an ambiguous social position, some adopting and benefiting from the disability label (e.g., gaining protection through anti-discrimination legislation and access to welfare provision) and others sensing stigma if identifying as a disabled person. The questions of what a (politicized) 'disability' identity is and can offer to people with bodies of difference and, further, the formation and role of organizations of disabled people, remain open.

Future Directions

Like all good research, the chapters in this book raise as many questions as they answer. By way of conclusion, we wish to suggest a number of related avenues for future research that emerge from the contributions to this volume. First, there is much to be gained from exploring the intersections and constructive tensions between health and disability research. Within geography, recent work on chronic physical illness and mental health has already necessitated some useful engagements. More broadly, the move towards embodiment has encouraged some intellectual exchange, but as Dyck (Chapter 14) reminds us, there is much that remains to be done. As Power (Chapter 6) suggests, exploring the tensions inherent within geographies of care and support from health and disability perspectives has the potential to elucidate complex interdependencies in the lives of both disabled and non-disabled people. More broadly, there is a need to recognize that notions of health and ill-health are critical to a fuller understanding of the complex geographies of disability at multiple scales. At the same time, disability scholars can make valuable contributions by questioning the continued conflation of illness, impairment and disability as a single (tragic) phenomenon in public health formulations (Gross and Hahn 2004).

Second, there continues to be an absence of work by geographers beyond Western and urban contexts, despite the fact that the majority of disabled people live in developing countries (Seelman 2001). Moreover, the continued absence of adequate nutrition, clean water, economic opportunity and basic health care likely account for up to half of the impairments experienced by the world's population

(Barnes and Mercer 2003). Geographers can make important contributions by producing empirical work beyond the West, and by forging linkages with researchers and organizations in developing contexts. Geographers can also broaden existing conceptual models of disability, producing 'a more holistic and flexible approach to understanding disability, with a greater focus on local and individual experience and on recognising the importance of geopolitical, social and cultural as well as economic contexts' (McEwan and Butler 2007: 463).

Third, geographers appear to be somewhat disengaged in relation to policy. As Edwards (Chapter 9) notes, it is not that geographers don't produce policy-relevant work. Rather the lack of influence may be a product of several factors including concerns about losing critical distance by entering policy debates, concerns about speaking *for* disabled populations in policy circles, and the fact that policymakers do not see geographers as a credible source for input (Imrie and Edwards 2007). Whatever the reason, there is a need for further discussion on this issue. Does policy-relevance, for example, necessarily mean working with and for governments, or are there other ways in which geographers might usefully engage? If we do wish to engage, what strategies can be used at varying spatial scales to facilitate these connections?

The policy issue also dovetails with another methodological question. The geographies of disability represented in this volume are almost entirely qualitative in nature. While qualitative methods have contributed much to geographies of disability and chronic illness, how might we broaden our understanding of such geographies with mixed method studies that incorporate a broader quantitative analysis of the social conditions of disabled people? We are aware of the shortcomings of quantitative data, not least their tendency to reify medically informed categories of impairment. However, this does not mean that they cannot be used critically and in combination with qualitative methods.

Fourth, there is considerable opportunity to critically engage with the geographies of 'able-bodiedness' and 'able-mindedness'. While there has long been recognition that the privileges of the able-body and mind are reproduced through the oppression of a disabled 'other', there have been few attempts to systematically unpack these 'able' categories. A number of questions might be posed. For example, how are able-bodiedness and able-mindedness produced as geographically and historically contingent constructs? What types of knowledge, practices and spaces are implicated in their reproduction? How might they be destabilized? Responding to these questions may also be critical for efforts to rethink notions of independence and dependence. Insights on how to proceed might be derived from feminist analyses of masculinity (e.g., Connell 1995).

Finally, we wonder about the extent to which disability scholarship has been integrated into, and helped to shape, a fully human geography. Gleeson (1999a) noted a decade ago that much of the geographical work on disability was being produced by a small number of social and cultural geographers, while areas such as economic, urban and historical geography remained largely unengaged. While geographers have produced valuable work on disability in the intervening years, we have a long way to go to make disability a central part of the discipline. The chapters in this volume demonstrate the important connections between disability research and the geographies of ageing (Wiles and Allen), feminist geographic scholarship (Longhurst) and political geography (Maddern and Stewart). Yet, there remains a pressing need to demonstrate the significance of disabled people's experiences to broader geographic scholarship on topics such as welfare state restructuring, immigration, social exclusion and cultural representation.

References

Barnes, C. 1991. *Disabled people in Britain and discrimination*. London: Hurst and Co.
Butler, R. 1994. Geography and the vision-impaired and blind populations. *Transactions of the Institute of British Geographers*, NS19, 366-368.
Butler, R. and Bowlby, S. 1997. Bodies and spaces: an exploration of disabled people's experiences of public space. *Environment and Planning D: Society and Space*, 15, 411-433.
Butler, R. and Parr, H. (eds) 1999. *Mind and body spaces: geographies of illness, impairment and disability*. London: Routledge.
Cameron, A. 2005. Geographies of welfare and exclusion: initial report. *Progress in Human Geography*, 29(2), 194-203.
Chouinard, V. 1997 Making space for disabling differences: challenges ableist geographies. *Environment and Planning D: Society and Space*, 15(4), 279-87.
Chouinard, V. and Crooks, V.A. 2005. 'Because *they* have all the power and I have none': State restructuring of income and employment supports and disabled women's lives in Ontario, Canada. *Disability and Society*, 20(1), 19-32.
Colls, R. 2006. Outsize/outside: bodily bignesses and the emotional experiences of British women shopping for clothes. *Gender Place and Culture*, 13, 529-545.
Connell, R. 1995. *Masculinities*. Berkeley: University of California Press
Cooper, C. 1997. Can a fat woman call herself disabled? *Disability and Society*, 12(1), 31-41.
Crooks, V., Dorn, M. and Wilton, R. 2008. Emerging scholarship in the geographies of disability. *Health and Place*, 14(4), 883-888.
Crooks, V.A. 2007. Exploring the altered daily geographies and lifeworlds of women living with fibromyalgia syndrome: a mixed-method approach. *Social Science and Medicine*, 64(3), 577-588.
Davidson, J., Bondi, L. and Smith, M. (eds) 2005. *Emotional geographies*. Aldershot: Ashgate.
Dean, K. and James, H. 1981. Social factors and admission to psychiatric hospital: schizophrenia in Plymouth. *Transactions of the Institute of British Geographers*, NS6, 39-52.

Dear, M. and Taylor, S. 1982. *Not on our street: community attitudes to mental health care*. London: Pion.

Dear, M. and Wolch, J. 1987. *Landscapes of despair: from deinstitutionalisation to homelessness*. Princeton, NJ: Princeton University Press.

Del Casino, V. 2001. Enabling geographies? Non-Governmental Organizations and the empowerment of people living with HIV and AIDS. *Disability Studies Quarterly*, 21(4), 19-29.

Driedger, M., Crooks, V. and Bennett, D. 2004. Engaging in the disablement process over space and time: narratives of persons with multiple sclerosis. *The Canadian Geographer*, 48(2), 119-136.

Dyck, I. 1995. Hidden geographies: the changing lifeworlds of women with multiple sclerosis. *Social Science and Medicine*, 40(3), 307-320.

Finkelstein, V. 1980. *Attitudes and disabled people*. New York: World Rehabilitation Fund.

Gant, R. and Smith, J. 1988. Journey patterns of the elderly and disabled in the Cotswolds: a spatial analysis. *Social Science and Medicine*, 27, 173-180.

Giggs, J. 1973. The distribution of schizophrenics in Nottingham. *Transactions of the Institute of British Geographers*, 59, 55-76.

Giggs, J. 1986. Mental disorders and ecological structure in Nottingham. *Social Science and Medicine*, 23, 945-961.

Gleeson, B. 1996. A geography for disabled people? *Transactions of the Institute of British Geographers*, NS21, 387-396.

Gleeson, B. 1999a. *Geographies of disability*. London: Routledge.

Gleeson, B. 1999b. Can technology overcome the disabling city, in *Mind and body spaces: geographies of illness, impairment and disability*, edited by R. Butler and H. Parr. New York: Routledge, 98-118.

Golledge, R. 1993. Geography and the disabled: a survey with special reference to vision impaired and blind populations. *Transactions of the Institute of British Geographers*, NS18, 1, 63-85.

Golledge, R., Loomis, J., Flury, A. and Yang, X. 1991. Designing a personal guidance system to aid navigation without sight: progress on the GIS component. *International Journal of Geographical Information Systems*, 5(3), 373-395.

Gross, B. and Hahn, H. 2004. Developing issues in the classification of mental and physical disabilities. *Journal of Disability Policy Studies*, 15, 130-134.

Hahn, H. 1986. Disability and the urban environment: a perspective on Los Angeles. *Environment and Planning D*, 4, 273-288.

Häkli, J. 2007. Biometric identities. *Progress in Human Geography*, 31(2), 139-141.

Hall, E. 1999. Workspaces: refiguring the disability–impairment debate, in *Mind and body spaces: geographies of illness, impairment and disability*, edited by R. Butler and H. Parr. London: Routledge, 138-154.

Hall, E. 2000. 'Blood, brain and bones': taking the body seriously in the geography of health and impairment. *Area*, 32(1), 21-29.

Hall, E. 2004. Social geographies of learning disability: narratives of exclusion and inclusion. *Area*, 36(3), 298-306.

Hall, E. and Kearns, R. 2001. Making space for the 'intellectual' in geographies of disability. *Health and Place*, 7, 237-246.

Hansen, N. and Philo, C. 2007. The normality of doing things differently: bodies, spaces and disability geography. *Tidjschrift voor Economische en Sociale Geografie*, 98(4), 493-506.

Holt, L. 2004. Children with mind-body differences and performing (dis)ability in classroom micro-spaces. *Children's Geographies*, 2(2), 219-236.

Hubbard, R. 2006. Abortion and disability: who should and who should not inhabit the world? in *The Disability Studies Reader* (2nd Edition), edited by L. Davis. New York: Routledge, 93-103.

Hughes, B. and Paterson, K. 1997. The social model of disability and the disappearing body: towards a sociology of impairment. *Disability and Society*, 12(3), 325-340.

Imrie, R. 1996a. Ableist geographies, disableist spaces: towards a reconstruction of Golledge's 'Geography and the disabled.' *Transactions of the Institute of British Geographers*, NS21, 397-403.

Imrie, R. 1996b. *Disability and the city: international perspectives*. London: Paul Chapman.

Imrie, R. 2006. *Accessible housing: quality, disability and design*. London: Routledge.

Imrie, R. and Edwards, C. 2007. The geographies of disability: reflections on the development of a sub-discipline. *Geography Compass*, 1(3), 623-640.

Ingram, A. 2008. Domopolitics and disease: HIV/AIDS, immigration and asylum in the UK. *Environment and Planning D*, 26, 875-894.

Kitchin, R. 2001. Using participatory action research approaches in geographical studies of disability: some reflections. *Disability Studies Quarterly*, 21(4), 61-69.

Laws, G. 1994. Oppression, knowledge and the built environment. *Political Geography*, 13(1), 7-32.

Lovett, A. and Gatrell, A. 1988. The geography of spina bifida in England and Wales. *Transactions of the Institute of British Geographers*, NS13, 288-302.

Mayer, D. 1981. Geographical clues about multiple sclerosis. *Annals of the Association of American Geographers*, 71(1), 28-39.

McEwan, C. and Butler, R. 2007. Disability and development: different models, different places. *Geography Compass*, 1(3), 448-466.

Milligan, C. 2004. Caring for older people in New Zealand: informal carers' experiences of the transition of care from the home to residential care. [Online: University of Lancaster, Institute for Health Research]. Available at www.lancs.ac.uk/users/ihr/staff/christinemilligan.htm.

Mohan, J. 2003. Geography and social policy: spatial divisions of welfare. *Progress in Human Geography*, 27(3), 363-374.

Morris, J. 1991. *Pride against prejudice*. London: The Women's Press.

Oliver, M. 1990. *The politics of disablement*. Basingstoke: Macmillan.

Pain, R. 2006. Social geography: seven deadly myths in policy research. *Progress in Human Geography*, 30(2), 250-259.

Papadimitriou, C. 2008. Becoming en-wheeled: the situated accomplishment of re-embodiment as a wheelchair user after spinal cord injury. *Disability and Society*, 23(7), 691-704.

Park, D., Radford, J. and Vickers, M. 1998. Disability studies in human geography. *Progress in Human Geography*, 22(2), 208-233.

Parr, H. and Butler, R. 1999. New geographies of illness, impairment and disability, in *Mind and body spaces: geographies of illness, impairment and disability*, edited by R. Butler and H. Parr. London: Routledge, 1-24.

Shakespeare, T. 2005a. Disability, genetics and social justice. *Social Policy and Society*, 4, 87-95.

Shakespeare, T. 2005b. Disability studies today and tomorrow. *Sociology of health and illness*, 27, 138-148.

Shakespeare, T. 2006. *Disability Rights and Wrongs*, London and New York: Routledge.

Sheldon, A. 2004. Changing technology, in *Disabling Barriers – Enabling Environments* (2nd Edition), edited by J. Swain et al. London: Sage, 155-160.

Smith, M. and Davidson, J. 2006. 'It makes my skin crawl…': The embodiment of disgust in phobias of 'Nature'. *Body and Society*, 12(1), 43-67.

Ward, K. 2005. Geography and public policy: a recent history of 'policy relevance'. *Progress in Human Geography*, 29(3), 310-319.

Wilson, J. 2007. (Re)writing the genetic body-text, in *The Disability Studies Reader* (2nd Edition), edited by L. Davis. New York: Routledge, 67-77.

Wilton, R. 2004. From flexibility to accommodation: workers with disabilities and the reinvention of paid work. *Transactions, Institute of British Geographers* NS29, 420-432.

Wolch, J. and Philo, C. 2000. From distributions of deviance to definitions of difference: past and future mental health geographies. *Health and Place*, 6(2), 137-157.

Zitzelsberger, H. 2005. (In)visibility: accounts of embodiment of women with physical disabilities and differences. *Disability and Society*, 20(4), 389-403.

Chapter 2

Disability, Embodiment and the Meaning of the Home

Rob Imrie

Introduction

> Empowerment is often found in the details of the mundane world. It comes from
> controlling access to personal space, from being able to alter one's environment
> and select one's daily routine, and from having personal space that reflects and
> upholds one's identity and interests (Ridgway et al. 1994).

A person's mental and physical well being is related to many circumstances, not the
least of which is the quality of their dwelling and home environment. An important
part of such quality is physical design and layout, and how far it enables the ease
of people's mobility and movement around the dwelling and the use of different
rooms and their facilities. For many disabled people, particularly those with
mobility impairments and/or who are dependent on the use of a wheelchair, the
design of dwellings is often not well suited to their needs (Borsay 1986, Heywood
et al. 2002, Imrie 2003, Karn and Sheridan 1994). Harrison with Davis (2000:
115), for instance, notes that the poor physical design of housing can prevent self-
management of impairments, 'and may exacerbate a condition'. Likewise, Oldman
and Beresford (2000: 439) suggest that the home lives of children with limited
mobility lack 'spontaneity' because they 'rely on an adult to move them around'.

   Such studies, amongst others, seem to indicate that disabled peoples' domestic
experiences are, potentially, at odds with the (ideal) conceptions of the home as
a haven, or a place of privacy, security, independence, and control. In part, this is
because design conceptions, in relation to floor plans and allocation of functions
to specific spaces, do not conceive of impairment, disease, and illness as part of
domestic habitation or being. The impaired body is rarely imagined or drawn into
domestic design and the production of dwellings or buildings more generally (see
Imrie 2003, Imrie and Hall 2001). This is unsurprising given that representations of
idealized domestic life revolve around what Hockey (1999) refers to as positively
perceived values, such as companionship and freedom, but tend to exclude, even
deny, other aspects of domestic life, such as disease, impairment and dying. This
reflects a broader problem with debates about the meaning of the home, and
housing studies more generally, in which the impaired body is rarely a subject of
comment and analysis.

In seeking to redress this, I divide the paper into three. First, I outline, in brief, some of the broader debates about the meaning of the home, and suggest that these are far from helpful in enabling a coherent understanding of disabled peoples' experiences of domestic environments. Second, I refer to testimonies from interviews with disabled people that highlight some of the paradoxes and problems of idealized conceptions of the home that hinge, in part, on little or no recognition of impaired corporeality as a potential part of home life. I conclude by developing the idea that there is an urgent need to 'corporealize' the meaning of the home, in which conceptions of domestic life become underpinned by an understanding of the interactions between a person's bodily or physiological condition, and their patterns of behaviour in the domestic environment.

Seeking to Embody the Meaning of the Home

It has been well established in housing studies that the home is one of the fundamental places that gives shape and meaning to people's everyday lives (Dupuis and Thorns 1996, Gurney 1990, Rakoff 1977, Saunders 1989, 1990, Saunders and Williams 1988). A burgeoning literature has, in various ways, explored the social, health, and psychological effects of the home (Allan and Crow 1989, Gilman 2002, Hopton and Hunt 1996, Madigan and Munro 1999). For instance, Sixsmith and Sixsmith (1991) note that the home is a symbol of oneself or a powerful extension of the psyche. It is a context for social and mental well being or, as Lewin (2001) suggests, a place to engender social psychological and cultural security. For others, the home is the focus for personal control and a place that permits people to fashion in their own image (Saunders 1990). In this sense, the domestic setting is, for Lewin (2001), a mirror of personal views and values (also, see Cooper 1995).

Gilman's (1903; reprinted 2002: 3) seminal text suggests that the home, ideally, should offer a combination of rest, peace, quiet, comfort, health, and be a place for personal expression. Indeed, throughout the twentieth century, the home has been counterpoised to work, as a place of retreat, social stability, and domestic bliss far from the travails of everyday life (see, for example, arguments in Rakoff 1977, Saunders 1990). From builders' marketing brochures that seek to sell the dream of the ideal home, to television programmes about selling a place in the sun, the home is popularly portrayed as the focus of convivial social relationships and a source of human contentment. It is, first and foremost, a place for family interaction and the setting for personal seclusion and intimate behaviour free from public comment or restraint. The home is also the setting for the development of personal values, and patterns of socialization and social reproduction more generally.

These characterizations of the home, however, do little to reveal the complexity of the cross cutting variables that imbue domestic space with meaning. Saunders and Williams (1988) and Saunders (1989, 1990) have been accredited with (re)igniting debates in housing studies about the meaning of the home that, in part, have gone some way to identifying such complexity and in fleshing out, empirically, what Lewin (2001) refers to as the home as a composite concept (also, see Allan and Crow 1989, Chapman and Hockey 1999). For Saunders and Williams (1988), the meaning of the home is not fixed but varies, potentially, between different household members, especially in terms of gender and age, and between households, especially in relation to differences in social class. They also suggest that people's experiences of, and meanings attributed to, the home may differ according to geographical context or setting.

Such studies indicate that the meaning of the home is unstable and transitory. Gilman (2002: 5) anticipated as much when, writing in 1903, she noted that 'this power of home-influence we cannot fail to see, but we have bowed to it in blind idolatry as one of unmixed beneficence.' For Gilman (2002: 8), despite the prevailing wisdom that homes were 'perfect and quite above suspicion', the home was a potential source of repression. In particular, she referred to women's exclusive confinement to the home as leading to 'mental myopia' in which the individual was made into 'less of a person.' Likewise, a range of feminist writers have sought to deconstruct ideal images of the home by suggesting that the home, for some women, is a place of captivity and isolation (Allan 1985, McDowell 1983). It is, as Goldsack (1999: 121) notes, 'less of a castle, and more of a cage.' Others note that the home is as much about the focus for the drudgery of domestic work as for personal pleasure, and a place of fear where, potentially, domestic violence takes place.

While these, and related, studies have done much to destabilize popular representations of the home, they tend to refer to abstract categories (e.g., gender, ethnicity, etc.) that rarely relate to, or reveal, how specific bodily or physiological phenomena interact with dwellings to produce personal experiences of, and generate particular meanings about, the home (although, for exceptions, see the excellent writings of Gurney 2000, nd). Indeed, I concur with Gurney (1990) who notes that it is problematical to explain the meaning of the home with reference only to generalized categories, such as class, income, or tenure. Rather, for Gurney, the significance of the home is influenced by different personal experiences. Foremost, I would contend, relates to the body in that, as Twigg (2002: 436) comments, the body is a necessary condition of life inasmuch that 'social life cannot proceed without this physiological substratum' (also, see Crossley 2001, Ellis 2000, Shilling 1993).

Others concur in noting that the body is the most significant referent of a person or, as Merleau-Ponty (1962: 150) notes, 'I am not in front of my body, I am in it, or rather I am it'. For Merleau-Ponty (1963: 5), the 'body is not in space like things; it inhabits or haunts space…through it we have access to space'. Here, the body, as a sensory and physiological entity, is constitutive of space or, as Lefebvre (1991: 174) comments, 'the most basic places and spatial indicators are first of all qualified by the body.' Physiological substratum is also core to domestic life in that the home is the focus for the care of the body, including washing, dressing, grooming, and preparation for entry to the world beyond the front door.

The physical design of dwellings is 'thoroughly embodied' in that each part of the domestic environment can be thought of as a 'body zone', or where particular bodily functions, both physical and mental, are attended to. Thus, the bathroom is the place for washing the body, while the bedroom is the place for physical and mental recuperation.

While such functional demarcations are neither inevitable nor unchangeable, they are part of a broader, and powerful, social and cultural encoding of what constitutes appropriate domestic space and their legitimate (bodily) uses. Such encoding, however, rarely relates to impairment, or to bodies that may require an integration of rooms and/or functions, or more flexible forms of domestic design. In particular, disabled people often experience the home as a series of 'disembodied spaces', or places that are designed in ways that are rarely attentive to their physiological and bodily needs and functions. Thus, interactions between features of bodily physiology, such as muscle wasting, and domestic design, such as heavy doors, can combine to demarcate domestic spaces that are off limits to (particular types of) impaired bodies. For Hockey (1999: 108), such embodied experiences, in which people are excluded from participation 'in the performance of home as idealized, is to undermine a view of home as a sanctuary or "place of secure retreat."'

Insights into disabled people's experiences of, and meanings associated with, the home, ought to proceed, however, by rejecting reductive conceptions of disability and impairment. Thus, the body is neither a naturalistic organic entity, unaffected by socialization, nor a socialized entity, unaffected by physiology. Rather, the body, and its interactions with domestic space, reflects a complex conjoining of physiological and social and cultural relations to produce specific, person-centred, meanings of the home. For instance, doorsteps have long been part of the aesthetic décor of dwellings, and reflect values about what constitutes appropriate design (see Milner and Madigan 2001). However, for wheelchair users, steps prevent ease of entry to homes. In such instances, the experience, and potential meaning, of the home, as a form of embodied encounter, is influenced by the interplay between physiological matter (i.e., the absence of use of limbs) and those social and cultural relations that give rise to, and legitimate, particular design features (i.e., steps).

In this sense, the paper makes the plea for an embodied understanding of the meaning of the home that, in the context of impairment, does not seek to explain such meanings purely in terms of biological phenomena (i.e., the medical conditions causes the experience and determines the meaning), or in terms that assert the primacy of social phenomena over biology (i.e., the organic or biological condition is irrelevant in the construction of people's experiences of, and associated meanings derived from, the home) (also, see Allen 2003, Oldman and Beresford 2000). Given these caveats, I now turn to an exploration of the multiple meanings of the home in relation to the experiences of people who are dependent, for most part, on the use of wheelchairs.

Disability, Domestic Design and the Home Environment

In investigating disabled people's feelings about disability and domestic design, two research methods were adopted. First, two focus groups were held in October 2002 with members of a disabled persons user group located in a UK south coast conurbation. This was followed, over the course of the next five months, by interviews with 20 individuals living in three different towns. Each person was contacted through the context of an intermediary, or gatekeeper. Thus, in one place, the chair of the local access group permitted me to give a lecture about my research to the group, and to leave my contact details for individuals to contact me if they wished to talk with me at a later date. In contrast, in the other places, I approached the chair of user groups, and they posted out my details to their membership inviting them to contact me if they were willing to talk about aspects of disability and domestic design.

Interviews normally occurred in the subject's home and were usually two to three hours in length. They covered a wide range of themes relating to individuals' life histories and, especially, their experiences of the home. Respondents talked about the various dwellings that they had lived in, and how parts of their domestic lives were affected by the onset of impairment. In particular, respondents were asked to articulate how the meaning of their home had been transformed, if at all, by the interaction between impairment and the physical design of their dwelling. The conversations were taped and transcribed and copies of testimonials were returned, for comment, to each individual.[1] The subjects are all individuals with various mobility impairments, ranging from those with problems of balance due to the early onset of Parkinson's disease, to individuals with advanced stages of multiple sclerosis that render them dependent, for some of the time, on a wheelchair.

The respondents live in a mixture of different types of dwellings including flats (five respondents), detached homes (four), institutional care settings (three), and terraced and semi-detached dwellings (eight). The mix of tenures is evenly divided between those occupying social (eight) and owner-occupied (nine) dwellings. Because the sample is 'self-selecting', and derived from access groups, the respondents are knowledgeable about design issues and were usually forthright in their opinions. Their membership of an access group meant that most were aware of the need to campaign, politically, to try and change the processes underpinning

---

1   Sending transcripts back to respondents is important in order to clarify points of detail or to give respondents the opportunity to change their story or qualify points that they have made. Some respondents did not reply and others were concerned about how they would 'come across' or 'sound'. They either deleted certain expressions or changed the wording of parts of the transcript. Respondents have been sent a draft copy of this paper for comment but no one has asked for any details to be taken out or re-written; all respondents have been presented anonymously and revealing details about them have not been included.

the production of disabling built environments. Most of the respondents, from their standpoint, had thought through the issues relating to design, disability, and domesticity and were able to articulate different ways in which they felt society should change. In a modest way, the transcripts are able to 'give voice' to the politicized knowledge of the respondents (see Letherby 2002 and Millen 1997 for an outline of these issues).

In describing and evaluating the research material, I divide the discussion into three. First, I develop and evaluate the argument that the design of home environments interact with impairment to produce, more often than not, a series of spaces that are rarely sensitized to the needs of disabled people. Rather, such spaces, so some claim, lead to, what Leder (1990) terms, 'corporeal dys-appearance' (also, see Paterson and Hughes 1999). Second, I consider how far, and in what ways, dominant representations of the ideal home, such as privacy, security, and sanctuary, accord with disabled people's experiences of their homes. Finally, I develop the argument that disabled people are not necessarily passive victims of insensitive domestic design but, depending on social, personal, and material resources, are able to influence aspects of the design and usability of the home environment.

Corporeal Dys-appearance and Privation in the Home

The physical design of housing tends to reflect a particular conception of corporeality based around a body that is not characterized by impairment, disease, and illness (Hockey 1999, Imrie 2003). For instance, most kitchen units in homes are provided as a standardized package in which tabletop and cupboard heights are reachable only by an upright person. People who are dependent on a wheelchair, or whose mobility is such that they have to hold onto a support structure to stabilize themselves, often find it impossible to use their kitchen unless it is adapted to meet their needs. Thus, as Ann recounted, about her kitchen before it was adapted 'It was too high, I couldn't have used the wheelchair, the cupboards were too high, the cooker was completely unusable, I would leave the thing on and oh, it just went on and on and on...As a mum it totally demoralised me'.

The design of most dwellings is also underpinned by values that rarely relate to, or incorporate, the needs of wheelchair users. Some respondents were angry that their homes were short of space to permit them ease of movement from one room to another, or even within rooms. For John, his bedroom is an apt example of where design values have been applied without relating to impairment. As he recalled:

> There are some basic assumptions. I'm just talking about a very simple basic thing like there is no way on this earth that my wheelchair can go to the other side of my bed. It doesn't matter what you do you can't configure the bedroom any other way, so the assumption must be that I'm not going to make my bed, that I don't need to get to the other side of the room.

John is unable to get access to the bedroom window and, consequently, he cannot open the window to air the room: He said that 'it's obviously assumed that I don't need to open my window'.

Others commented on the lack of space as the most important factor in preventing them from getting access to rooms and living as they please in their homes (also, see Oldman and Beresford 2000, Percival 2002). As Carol said:

> The kitchen is really very small and when you're maneuvering a wheelchair you do need a bit of space. You can hardly get your furniture in the lounge and you have to eat in it. It's things like this, and I'm thinking to myself, you've got a life and you want to lead your life and this isn't really helping you.

Similarly, Janet was unhappy with the shortage of space in her WC which, she felt, compromised the quality of her life:

> If my loo had been built eighteen inches longer it would've meant I could've got my whole wheelchair in, but as it is I can't use it with the wheelchair…I have to leave the door open, and it just brings it home to you about what you can't do in your own home.

Such examples serve to illustrate what Leder (1990: 84) refers to as the 'dys-appearing' body or where, as he suggests, 'the body appears as a thematic focus of attention but precisely in a dys-state' (also, see Paterson and Hughes 1999). What Leder (1990) is inferring is that, in everyday life, consciousness of the body, either by oneself or by others, is minimal or non-existent. That is, the body has, more or less, disappeared from consciousness. It only reappears, explicitly, in a context of pain, disease, or bodily dysfunction. Its reappearance is characterized by encounters with the embodied norms of everyday life, or those that are reflective of, primarily, non-impaired forms of carnality. Such norms serve to reproduce a world in the image of non-impaired bodies, with the consequence that, in Paterson and Hughes' (1999: 603) terms, the impaired body is experienced 'as-alien-being-in-the-world'.

This characterization was recognized, and understood, by a number of respondents who said that aspects of design quality in their homes are not sensitized to the needs of impaired bodies. For instance, Clare referred to the poor quality of internal walls that prevented her erecting a stair lift: 'it didn't even have a solid wall to put the stair lift in, you know, it was only like chipboard or something… I've had to live down here because of this.' Jim was also frustrated because he was unable to reach a lot of things in his (council-owned) home, and could only see the sky out of his windows. In noting this, he highlighted a common observation of respondents about window levels being too high to see outside, and that window design 'is just done for people who can stand up, not for me.' Carol felt that a similar situation in her home was akin to cutting off part of her life, 'because it's nice to be able to look out and see what's happening.'

These experiences are illustrative of what is not embodied in the design of dwellings, that is bodily difference and, in particular, impairment. Grosz (2001), in referring to sexual difference, suggests that part of the problem is that designers foreclose on the question of sexual difference and do not sensitize design to the sexualized nature of embodiment in the built environment; consequentially, buildings, in Grosz's interpretation, are phallocentric. So too in relation to impairment in which the interactions between many disabled people and the design of dwellings is akin to an absent presence. The body is simultaneously there but not there, characterized by material practices (i.e., moving from room to room, bathing, etc.) which draw attention to 'out-of-place' bodies, or bodies unable to operate wholly in environments characterized by the embodied norms of society.

Such embodied norms rarely recognize the intrinsic nature of impairment to the human condition, and were highlighted by respondents in relation to the separation of functions by floor levels. In particular, the spatial separation of the bathroom and toilet (upstairs) from daily living functions (downstairs) is premised on a walking and mobile person. For Elaine, the deterioration of her leg muscles restricted her to downstairs, with the consequence that she had difficulties bathing. As she said: when we moved in here four years ago my husband had to shower me in the back garden with a hose pipe, because I couldn't get upstairs and there wasn't even a shower downstairs, we put that in ourselves.' Likewise, a traumatic experience for Heather in her former home made her realize that upstairs was no longer feasible for her to use. As she recalled:

> It was difficult to use the house. The stairs was twisting and I had fallen down the stairs…I'd lost my balance, and I injured my ribs really bad. I was shouting out for help and my family didn't wake up (laughs) and I was in the most awful pain for a long time…after that, I never went upstairs again.

The best that respondents felt they could get was partial access to, and use of, some rooms. Thus, even when a dwelling has, supposedly, been designed to facilitate ease of use by wheelchair users, aspects of design do not always work well. For instance, Pete's house was, in his words, 'purpose built around me and it works'. However, he admitted that:

> I'm not independent in it, I can't pull my own curtains, I can't get into all the corners of my house, at all'. As Pete suggested, 'there's still parts of it. I will never use where other people will be able to use their homes, so it depends again on what you mean. Can I fully function within my house? Well, I can get in and out and my care could be provided within that facility, etc. but no, I can't use my home in the way I still think I would be able to, given the sort of facilities within it that are available to be put in.

In this respect, much of Pete's house is indicative of spaces of exclusion, or places that are not habitable in the present circumstances.

For most respondents, living in the home is achieved by accepting, and adapting to, the standards of design that reflect the primacy of non impaired bodies. While respondents often expressed anger about this, they felt that there were few options open to them. For instance, Joe commented on the unfairness of imposing on him domestic design that tended to amplify, and draw attention to, his impairment: 'if I try and use that room then it only shows up that my body isn't up to it…it's not me though, it's the lack of space in there'. However, he felt he had no option but to compromise, although he felt it was all one-sided in that disabled people are the ones who have to take what is on offer. As he said: 'you compromise all the time. I hear people all the time saying 'It's good, I can get by, I make do, I'm quite happy.' I don't hear that from temporarily able-bodied people. They're not saying that about their homes.'

The feelings of a state of body-out-of-place in the home were, more often than not, related to design details, or the micro-architecture, of the dwelling (also, see Imrie 2000). Thus, it was often the subtle aspects of the design of the home environment that caused most problems. For instance, John referred to the fitting of an electric window to permit ease of opening of windows by the use of remote control. However, as he said: 'I mean, my electric window is beautiful, wonderful, but the switch is on the pelmet (laughter) and out of reach. It's like when they fitted it they didn't look at me or ask me if it was OK. They just did it'. Likewise, Ann, until recently, was unable to transport food around the house due to a slight gradient in the floor:

> This floor level in here was two inches lower than the hall. So I couldn't bring food in and out, I wouldn't have been able to have used a tray because everything was going to slop…the detail was very minute and you couldn't see it, but it was very major to me.

Impairment and De-stabilizing the Meaning of the Home

Binns and Mars (1984: 664) suggest that the ideal of the home as sanctuary is undermined in circumstances where the home environment becomes 'the product of withdrawal from wider social networks.' Indeed, for some respondents, broader social, attitudinal, and environmental circumstances, beyond the immediate confines of their home, had led them to 'stay-at-home' and rarely venturing beyond the front door. For instance, Harry recounted demeaning reactions from 'friends' concerning his inability to access, unaided, stepped thresholds into their homes: 'they think I'm being awkward…it's not as friendly an atmosphere as what it used to be, when I was up and walking…people say I'm seeking attention or whatever. They're wrong about that (laughs)'. For Harry, it has become easier not to visit friends or to expose himself to possible ridicule or suggestions that 'he's putting it on'. Rather, as Harry said, 'I spend most time indoors, and it feels like I'm confined to quarters.'

In other instances, social interactions have been curtailed or have stopped altogether. As Harry noted, 'I don't get invited to some of the parties any more, as they've got to lift me into the house'. Such situations have also prompted Harry to withdraw 'voluntarily' from most social engagements because, as he said, 'it means I've got to rush back here just to use the toilet'. Others also recalled how their lives have become 'home-bound' because they cannot get into friends' homes. As Ann said, 'well, obviously you can't get into them. There isn't really a home that I can get into. Obviously I can get into Mum's, but as for anywhere else, it's not, you know, there isn't anyone that I can actually go and visit.' Likewise, Carol's recent dependence on using a wheelchair has changed, in part, her patterns of mobility. She is less likely to go outside the home to visit people. As she said, 'I can't go and visit my sister. Three steps up to her front door, to start with. My mum's the same. I can't go to my hairdresser to get my hair done. She has to come to me, for the same problem, access to her house'.

Others recounted similar tales of how the onset of impairment led to a form of entrapment in housing circumstances that were socially and psychologically damaging. For instance, John was more or less confined to his house for six months awaiting adaptations to be carried out by the local authority. During this period he was unable to get out of the front door unaided and, as he said, 'it's no good saying to me "Oh yes, John. Oh, we know you've got MS, but we'll get the house sorted out in six months' time." By that time the damage is done, you know, you're stuck indoors, you're not going out, you don't want to come out, you've lost your confidence'. For John, his confinement to the home had significant social and personal consequences. As he explained:

> If you're stuck indoors for a month, six weeks, a month, it's very hard to start getting back out, when you used to go round the pub and meet your mates or walk out in the garden and see people walking down the street. That confidence goes and that's part of the trouble, once it goes it takes twice as long to get it back.

For John, and other respondents, the home had become the place of confinement and, far from being a haven, was, in part, a signifier of a life that had been lost. Their testimonials confirm, in part, Allan and Crow's (1989: 4) observation that the experience of privacy in the home is not always positive and that it can 'signify deprivation as well as advantage.' Ann amplified this point:

> It makes friendships and relationships very difficult, because they always have to come to me, you know, you accept things, you know that things are going to change or they have changed and I've just accepted them really. It is just part of life now. Which is why it makes it all the more, sometimes it makes it harder because I have always had this vision that I would be out and about and doing, and now I'm not, it's not going to happen.

Likewise, Elaine said that because of her weakening muscles, physical impediments on the pavement and the lack of access into the local shops, she had stopped going out. For her, 'it all stops at the garden gate'.

Others recalled the loss of independence and personal control in their home due to interactions between impairment and physical design. The effect, for some, was the onset of a series of social and psychological traumas. As Trish explained:

> I used to live in a two-up and two-down house and then I got an impairment that made me dependent on a wheelchair. I couldn't get upstairs and I was reduced to living downstairs. The sanitary conditions were awful and I was depressed and could see no way out and my living situation was not supporting my needs.

For Trish, the home became associated with a complete loss of independence and the performance of personal acts in degrading situations. As she said:

> My husband used to carry me upstairs and there were so many practical issues. He had to get my dresses from upstairs and I had to use a bucket for a toilet and I had to be bathed on towels downstairs in the living room. The experience made me realise that the correlation between psychological and physical states should not be underestimated.

For others too, the home was less a place of independence and more a context in which things had to be done for them. Thus, everyday household activities became, with the onset of impairment, more or less impossible to do without some assistance. For instance, Judith, who lives with her parents, now depends on her mother to cook for her because the kitchen tabletops and cooker are too high for her to reach. Cupboard and storage space are beyond reach and the practical implications for Judith are such that she is unable to gain easy access to stored foodstuff: 'if I want a packet of noodle soup I can't get it, it's up there. So I've gotta go, 'Dad, can you pass me that, can you pass me that.' Her father added that 'there's got to be somebody in there, if nobody's in the house she's had it you know.' If her parents are away for the weekend, Judith has recourse to a microwave machine or the oven: 'I'd just live off jacket potatoes and cheese in the microwave, or like fish and chips in the oven.'

Moreover, the idea that the home might provide for personal privacy is not always the case. For instance, Ann is constrained in using her downstairs WC because there is no guarantee that she can use it without being seen by another family member. Although she lives on her own, family members and a carer come and go without warning and they have a key to the front door. As Ann explained: 'I can't take the wheelchair in, I have to stand, I can't close the door as the chair blocks it and the front door's there and everyone in the family has a key, so I mean, it's not ideal and anyone could come in at any time.' Others felt that the design of their homes was such that they would never be able to easily function as private individuals. As Carol suggested: 'I always get the feeling that they purposely built these places not always for somebody who lives on their own but for somebody who lives with somebody, so they don't do it all, or don't have a life, quite frankly.'

Such experiences were, for these respondents, destabilizing and left them feeling that they had little control over circumstances. This, then, suggests that the nature of privacy in the home is never stable or guaranteed and, as Allan and Crow (1989: 3) suggest, 'an individual's ability to secure some degree of privacy is conditional.' Likewise, the idea of the home as a retreat, haven, or place of sanctuary and security is not always borne out, particularly in a context where a deteriorating body requires third party care and attention. For instance, some respondents felt that care staff were 'invasive' and made them feel uncomfortable and insecure. Ann recalled that it was difficult to find good carers locally because of high demand and the unpopularity of the job. While she praised the good ones that had cared for her in the past, recent experiences had been a problem. As she said: 'the ones I've had recently have been indifferent to me, they just saw the wheels and this thick person that's not got a brain.'

For Ann, these individuals were akin to having a stranger come into the home and then treat it in whatever way without permission:

> You had those that were literally there to earn some money and would sit here for three hours, and I couldn't get them to move, they just didn't want to do the work...they just did as they pleased, used my phone, cooker, the lot, and I would dread them coming.

Jim recalled similar situations with relatives who took it on themselves to come around unannounced to purportedly help out. Jim did not recognize his home as his haven because, as he said, 'I was always on edge, either waiting for them to turn up or for them to leave and sometimes they'd be here all day and just do as they pleased...it didn't matter what I'd say to them.' In such circumstances, the home, for Jim, felt like a place where he could not exercise much autonomy or control or close the door on outsiders.

Insecurity was also felt by those who said that they had attracted negative comment when outside, and did not want to draw attention to the fact that an impaired person lived in the house. For Harry: 'you want to blend in and not reveal that you can't walk. It makes you a target.' Others concurred and some respondents were wary about fitting a ramp up to their front door for fear of it labelling them as 'defective' and 'different.' As Carol said:

> I mean, I want to be able to live in my home but I don't want it to be screaming at anybody that walks in, to be inhibited because a disabled person lives here. That's the other thing, you know. I'm very, very conscious of this because one of my sons particularly found it very, very difficult to come to terms with it, and I don't want it screaming 'Oh dear, this poor woman lives on her own, she's in a wheelchair.'

This, then, illustrates, in part, the point by Saunders and Williams (1988) that the external physical features of the house convey subtle shades of meaning and act as signifiers to the outside world.

Resisting Domestic Design and Generating Usable Spaces

The social model of disability has tended to dominate research and writing on issues about disability and the built environment (Gleeson 1998, Oliver 1990). It posits that the design and layout of particular physical objects, such as steps or street furniture, is the primary source, or determinant, of a person's disability. Thus, insensitive design, based on values that do not recognize disabled people as users of buildings, creates physical barriers that prevent ease of mobility and access. This perspective, while not without merit, tends to conceive of individuals as 'victims' of circumstances beyond their control, in which they are oppressed by social and environmental factors. Impairment, and physiology, is more or less disregarded as having little role to play in determining disabled people's experiences of the environment. In Allen's (2000) terms, the human body is treated, potentially, as a 'physiological dope', without agency or the capacity to ameliorate or circumvent the 'given' conditions of existence.

In concurring with Allen (2000), I suggest that disabled people are not passive victims of insensitive design, nor necessarily resigned to dependence on others to facilitate aspects of their home lives. Far from it, the experiences of disabled people in this study, and elsewhere, illustrate the capacity to generate usable spaces out of the social and physical impediments that are placed in their way (Allen et al. 2002, Hawkesworth 2001, Heywood et al. 2002, Oldman and Beresford 2000, Percival 2002). For instance, Allen et al. (2002: 65) notes that parents of vision-impaired children do not necessarily see them as victims of the built environment. This is because most are able to construct, what Allen et al. (2002) refer to as, 'memory maps', or guides, of their home and neighbourhood environment that permit them to navigate, with relative ease, from one space to another.

For Harry, for example, the development of Parkinson disease has led to a re-appraisal of how best to use the house. As he said, 'well, I don't use upstairs often…there's no way I can actually do the stairs.' Harry did what many do by moving downstairs permanently. He set about changing the layout, taking out doors, installing external ramps, front and rear, removing the carpet, installing a new shower unit, and fitting lever taps. Anything he needs to use in the kitchen has 'been brought down to ground level' and, as Harry said, 'it's like a new home now, set up easily for me to use.' Judith's parents did likewise by building an extension on the ground floor that serves as an en-suite bedroom for their daughter and an accessible ramp to give her ease of access from the back door to the street (see Figure 2.1). As Judith's father said: 'it's been quite a major civil engineering task to adapt this house…she has an accessible way in and her own place to do as she wants'.

Figure 2.1   Getting out of the house

Note: The photograph shows Judith at the back door of her home at the top of a purpose built ramp. It has a shallow slope that makes it easy for her to get in and out of the house from the street without any need for assistance

The strategies deployed by respondents were, in part, dependent on income and social class. For respondents on low incomes, and living in council or housing association property, it was often a struggle to get things changed (see Heywood et al. 2002). As Jenny observed:

> If you've got no income and Social Services are making the alterations for you, you will have had a fight that's probably gone on three or four years to get it, and the chances of you succeeding again getting it if you move to another house is not very high, so you never want to move, you stay where you are.

Others concurred in expressing their frustration with delays in getting adaptations done. As Stan noted, 'this is one of the arguments I've had with Social Services for years and years and years. If you need handrails and a ramp, or a toilet adapted, you need it quickly…you know, when you're disabled you need help quickly'.

In contrast, those with higher incomes, and who owned their home, had more choice about how and when to adapt the domestic environment. Jenny expressed a common view: 'If you're middle class and you can afford to do it in the manner that I have, and you've got an income, you're earning money…then you do it'. Jenny's income had given her the means to install a state of the art kitchen and knock down internal walls to create more space (see Figure 2.2). For Jenny, what was important was not a singular or generalizeble approach to adapting her home. Rather, as she said, 'you want to bespoke what is done very precisely for yourself. You won't really want somebody to say "Here's a wheelchair ramp, you make do with that."' In this regard, Jenny had spent thousands of pounds in changing internal layout, and creating the sort of spaces that would permit her access to, and use of, all the rooms in the dwelling.

Respondents, regardless of income or tenure, were able to rearrange layout by, primarily, 'clearing up the clutter' and making space to facilitate ease of movement and use of rooms. Jenny moved into her house when she could walk and furnished it throughout. As she said: 'the house was designed for no more than a walking disabled person…and now it's inconvenient for me.' However, her more or less constant use of a wheelchair now means that 'if I wanted to get into a room I have to push chairs out of the way to get to the far wall…there was furniture everywhere.' For Jenny, the solution was to sell the furniture, or, as she said: 'I've just chucked everything out and we're now in a situation where there's not even any chairs for anybody to sit on.' Others have done similar things and Heather, living in housing association property, 'got rid of the big furniture and put up grab rails everywhere.'

Like Heather, other respondents have changed aspects of the micro architecture of their homes that had previously made a big difference to their mobility around the home. For Carol, the floor surface had to be changed when she became dependent on a wheelchair. As she said: 'I had carpet everywhere when I first moved in as I could still walk although not very well. Because when you're able to walk your feet skid, and you need carpet.' However, the carpet had prevented ease

Figure 2.2   An accessible domestic space

Note: Jenny's kitchen is customised to her needs. It permits her to get her knees under the work surfaces and close to food preparation areas. All fixtures are easy to reach and designed around the dimensions of her wheelchair. She can also reach the window and let fresh air into her home.

of movement of her wheelchair and so she ripped it up to reveal wooden floors which she restored. Likewise, Jim persuaded the local authority to provide a grant to adapt the downstairs toilet door, so that it now slides open and permits easier access than was hitherto the case. However, it was a struggle to get this: 'it's a big issue, apparently you're not supposed to have a sliding door in a bathroom, don't ask me why, that's what I was told.'

Adapting and/or the reorganization of the physical layout of domestic space are not the only means for disabled people to exercise some control over the use of the home environment. Some research suggests that energy saving strategies are part of the daily routine of people with particular types of impairment (Oldman and Beresford 2000, Percival 2002, Rubinstein 1989). For Carol, home life has always revolved around preserving her bodily energy and organizing tasks in ways that enable her to get through the day. As she commented:

> What you're doing is you're parcelling up energy when you're disabled. I've been doing this since I was seven. You know how much energy you've got for the day and you know what you're using on each of the tasks, and there's a whole list of tasks that you're buggered if you're gonna waste any time on. One

is washing up, eating is another one. When you're really tired you do not cook. You stand at the work top, you stand at the fridge and you eat, you know, a bit of cheese and a tomato and that's it.

Jenny followed a similar routine and was able to use her income to hire people to do some of the basic household chores. As she said:

When I moved into my house there was no question of me not having a cleaner for a few hours a week, just to do things like the floors and the dusting and the washing, she did my washing. And she does the ironing, which was the critical thing, the sheets and the ironing; I'll do the other things.

For most respondents the need to think ahead is paramount because of the knowledge that bodily deterioration will necessitate different ways of using the home. Jenny bought her present house when she was able to walk without the use of a wheelchair, but knew that, in time, she would be dependent on a wheelchair to get around. As she explained:

I bought the house because it was very flat. I was actually going round saying to my relatives 'Oh, this will do for a wheelchair,' and they were going 'Don't be stupid, you'll never go in a wheelchair.' And I always sort of knew my limitations; I knew it was on the cards.

Carol, who lives in council-owned property, pre-empted a move to more suitable accommodation by discovering an empty mobility house near to where she lived. As she said: 'I had a tip-off that it was empty, got in touch with the council, told them what I needed, and now a year later they've gutted it and made it up to what I need.'

However, such behaviour and/or actions appear to be no more than 'little victories' in a context whereby the design of most homes remains resistant to the needs of impaired bodies. Indeed, respondents were of the view that the only way to (re)claim domestic space for impairment is if professional experts, such as builders, architects, and occupational therapists, respond to experiential information and guidance provided by disabled people themselves (also, see Imrie and Hall 2001, Turner 1976). However, too often, as Jim said, 'the assumption, I presume, is that you've got no skill or knowledge that would actually have been of value to the person putting in your door'. Others concurred, with Jane expressing her frustration at the attitudes of the builders who had adapted her house: 'that's the interesting assumption about disabled people, isn't it, we obviously have got nothing to contribute back.'

Conclusions

The testimonials in this paper suggest that there are tensions between ideal conceptions of the home and the material, lived, domestic realities of disabled people. While aspects of the home may well provide for privacy, sanctuary, security, and other aspects of 'ideal' domestic habitation, such provisions are always conditional, contingent, never secure, and likely to be challenged by, amongst other things, the onset and development of bodily impairment. However, explorations of the meaning of the home, and housing studies more generally, rarely consider the body and impairment and its interactions with domestic space. This is curious because impairment is a significant, and intrinsic, condition of human existence and can affect anyone at any time (see Bickenbach 1993, Marks 1999, Zola 1989). In this sense, a person's feelings about, and experiences of, the home cannot be dissociated from their corporeality, or the organic matter and material of the body.

Indeed, dominant representations of the meaning of the home, propagated by builders, architects, and others, are underpinned by specific conceptions of embodied domestic spaces that do little to acknowledge the possibilities of bodily impairment as part of domestic habitation (see, for instance, Hockey 1999, Imrie 2003, Imrie and Hall 2001, Milner and Madigan 2001). Such representations revolve around the home as part of the ideal of family life, in which non-impaired bodies, with relative independence of movement and mobility, are paramount. The dominance of non-impaired carnality is reflected in physical design that, as the testimonials suggested, rarely include the fixtures, fittings, or spaces to enable the ease of use of domestic spaces by disabled people. Rather, such spaces, for many disabled people, are potentially disembodying in the sense that they deny the presence or possibility of bodily impairment and, as a consequence, likely to reduce the quality of their home life.

The interaction between domestic design and impairment produces a complexity of bodily experiences that ought to be the basis for the development of debates about the nature and substance of housing quality. However, too often debates about housing quality are about physical design and technical standards per se, in abstraction from bodily, or organic, form and function (also, see Harrison 2003). Such debates are themselves 'disembodied' in failing to refer to, or specify, the human context or content that comprises domestic habitation. An example of this is Part M of the building regulations that require builders to construct new dwellings to minimum standards of accessibility to facilitate ease of entry for wheelchair users. The regulations are deficient because the standards refer to disabled people in abstract and generalized terms, such as 'people who have an impairment which limits their ability to walk' (Department of the Environment, Transport and the Regions 1999: 5).

Such definitions reduce corporeal complexity to something that is limited, even fixed, in type and scope, and that can be accommodated within a narrow range of physical or technical responses. However, as the data in this paper indicate, bodily impairment is neither fixed nor static, or something that acquires meaning or function independent from social context or setting. Rather, as respondents noted, their home lives revolved around resolving issues relating to functioning in restrictive spaces, in contexts whereby bodily changes, particularly organic deterioration, were manifest realities. Housing quality, then, cannot be understood, or defined, separately from an understanding of the interactions between organic matter and the domestic setting, of which physical design is a component part. This should be one focus for seeking to develop an approach to housing studies that recognizes the importance of embodiment in influencing people's experiences of, and meanings attributed to, the home.

Acknowledgements

My thanks to the Economic and Social Research Council (grant number R-000-23-9210) who funded the research that this chapter is based on. I am particularly grateful to the research participants for giving up their time to share their experiences of the home with me. I would also like to thank Marian Hawkesworth for providing pointed comments and observations on an earlier draft of this chapter.

References

Allan, G. 1985. *Family Life: Domestic Roles and Social Organization.* Oxford: Blackwell.

Allan, G. and Crow, G. (eds) 1989. *Home and Family: Creating the Domestic Sphere*. London: Allen and Unwin.

Allen, C. 2000. On the 'physiological dope' problematic in housing and illness research: towards a critical realism of home and health. *Housing, Theory and Society*, 17(1), 49-67.

Allen, C. 2003. On the socio-spatial worlds of visual impaired children, or, Merleau-Ponty + Bourdieu = the socio-spatiality of the habitus. *Urban Studies*, 41(3), 487-506.

Allen, C., Milner, J. and Price, D. 2002. *Home is Where the Start is: The Housing and Urban Experiences of Visual Impaired Children*. York: Joseph Rowntree Foundation.

Bickenbach, J. 1993. *Physical Disability and Social Policy*. Toronto: University of Toronto Press.

Binns, D. and Mars, G. 1984. Family, community and unemployment: a study in change. *Sociological Review*, (32), 662-695.

Borsay, A. 1986. *Disabled People In The Community: A Study of Housing, Health and Welfare Services*. London: Bedford Square Press.

Chapman, T. and Hockey, J. (eds) 1999. *Ideal Homes? Social Change and Domestic Life*. London: Routledge.

Cooper Marcus, C. 1995. *Houses as a Mirror of Self*. Berkerley: Conan Press.

Crossley, N. 2001. *The Social Body: Habit, Identity, and Desire*. London: Sage Publications, London.

Department of the Environment, Transport and the Regions. 1999. *Approved Document M: Access and Facilities for Disabled People*. London: DETR.

Dupuis, A. and Thorns, D. 1996. Meanings of home for the older home owner. *Housing Studies*, 11, (4), 485-501.

Ellis, K. 2000. The care of the body, in *Social Policy and the Body: Transitions in Corporeal Discourse*, edited by K. Ellis, and H. Dean. Basingstoke: Macmillan, 23-44.

Gilman, C. 2002. *The Home: Its Work and Influence*. London: Alta Mira Press.

Gleeson, B. 1999. *Geographies of Disability*, London: Routledge.

Goldsack, L. 1999. A haven in a heartless world? Women and domestic violence, in *Ideal Homes? Social Change and Domestic Life*, edited by T. Chapman and J. Hockey. London: Routledge, 121-132.

Grosz, E. 2001. *Architecture from the Outside: Essays on Virtual and Real Space*. Massachusetts: MIT Press.

Gurney, C. 1990. The meaning of the home in the decade of owner occupation. *School for Advanced Urban Studies Working Paper 88*. Bristol: University of Bristol, SAUS Publications.

Gurney, C. 2000. Transgressing private-public boundaries in the home: a sociological analysis of the coital noise taboo. *Venereology*, 13, 39-46.

Gurney, C. undated. 'The neighbours didn't dare complain': some taboo thoughts on the regulation of noisy bodies and the disembodied housing imagination. Available at: www.cf.ac.uk/uwcc/cplan/enhr/files/Gurney-C2.html.

Harrison, M. 2003. Towards homes that facilitate: considering physical housing quality issues in diverse social settings. Paper presented at the ESRC Workshop *Housing Quality, Disability, and Design*, Senate House, University of London, 3 June.

Harrison, M. with Davis, C. 2000. *Housing, Social Policy and Difference: Disability, Ethnicity, Gender and Housing*. Bristol: Policy Press.

Hawkesworth, M. 2001. Disabling spatialities and the regulation of a visible secret. *Urban Studies*, 38(2), 299-318.

Heywood, F. Oldman, C. and Means, R. 2002. *Housing and Home in Later Life*. Milton Keynes: Open University Press.

Hockey, J. 1999. The ideal of home, in *Ideal Homes? Social Change and Domestic Life*, edited by T. Chapman and J. Hockey. London: Routledge, 108-118.

Hopton, J. and Hunt, S. 1996. The health effects of improvement to housing: a longitudinal study. *Housing Studies*, 11(2), 271-286.

Imrie, R. 2000. Disability and discourses of mobility and movement. *Environment and Planning A*, 32(9), 1641-1656.

Imrie, R. 2003. Housing quality and the provision of accessible homes. *Housing Studies*, 18(3), 395-416.

Imrie, R. and Hall, P. 2001. *Inclusive Design: Developing and Designing Accessible Environments*. London: Spon Press.

Karn, V. and Sheridan, L. 1994. *New Housing in the 1990's: A Study of Design, Space and Amenity in Housing Association and Private Sector Production*. York: Joseph Rowntree Foundation.

Leder, D. 1990. *The Absent Body*. Chicago: Chicago University Press.

Lefebvre, H. 1991. *The Production of Space*. Oxford: Blackwell.

Letherby, G. 2002. Claims and disclaimers: knowledge, reflexivity and representation in feminist research. *Sociological Research Online*, 6(4). Available at www.socresonline.org.uk.

Lewin, F. 2001. The meaning of home amongst elderly immigrants: directions for future research and theoretical development. *Housing Studies*, 16(3), 353-370.

Madigan, R. and Munro, M. 1999. 'The more we are together': domestic space, gender and privacy, in *Ideal Homes? Social Change and Domestic Life*, edited by T. Chapman, and J. Hockey, London: Routledge, 61-72.

Marks, D. 1999. *Disability: Controversial Debates and Psychosocial Perspectives*. London: Routledge.

McDowell, L. 1983. Towards an understanding of the gender divisions of urban space. *Environment and Planning D*, 1(1), 59-72.

Merleau-Ponty, M. 1962. *The Phenomenology of Perception*. London: Routledge and Kegan Paul.

Merleau-Ponty, M. 1963. *The Primacy of Perception*. Evanston: Northwestern University Press.

Millen, D. 1997. Some methodological and epistemological issues raised by doing feminist research on non-feminist women. *Sociological Research Online*, 2(3). Available at www.socresonline.org.uk.

Milner, J. and Madigan, R. 2001. The politics of accessible housing in the UK, in *Inclusive Housing in an Ageing Society: Innovative Approaches*, edited by S. Peace and C. Holland. Bristol: The Policy Press, 77-101.

Oldman, C. and Beresford, B. 2000. Home sick home: using the housing experience of disabled children to suggest a new theoretical framework. *Housing Studies*, 15(3), 429-442.

Oliver, M. 1990. *The Politics of Disablement*, Basingstoke: Macmillan.

Paterson, K. and Hughes, B. 1999. Disability studies and phenomenology: the carnal politics of everyday life. *Disability and Society*, 14(5), 597-611.

Percival, J. 2002. Domestic spaces: uses and meanings in the daily lives of older people. *Ageing and Society*, 22, 729-749.

Rakoff, R. 1977. Ideology in everyday life: the meaning of the home. *Politics and Society*, 7(1), 85-104.

Ridgway, P., Simpson, A., Wittman, F. and Wheeler, G. 1994. Home making and community building: notes on empowerment and place. *Journal of Mental Health Administration*, 21(4), 407-418.

Rubinstein, R. 1989. The home environments of older people: a description of the psychosocial processes linking person to place. *Journal of Gerontology*, 44, 45-53.

Saunders, P. 1989. The meaning of 'home' in contemporary English culture. *Housing Studies*, 4(3), 177-192.

Saunders, P. 1990. *A Nation of Home Owners*. London: Unwin Hyman.

Saunders, P. and Williams, P. 1988. The constitution of the home: towards a research agenda. *Housing Studies*, 3(2), 81-93.

Shilling, C. 1993. *The Body and Social Theory*. London: Sage Publications.

Sixsmith, A. and Sixsmith, J. 1991. Transitions in home experience in later life. *Journal of Architectural and Planning Research*, 8(3), 181-191.

Turner, J. 1976. *Housing by People: towards autonomy in building environments*. London: Marion Boyars.

Twigg, J. 2002. The body in social policy: mapping a territory. *Journal of Social Policy*, 31(3), 421-439.

Zola, I. 1989. Towards the necessary universalising of a disability policy. *The Milbank Quarterly*, 67(2), 401-428.

Chapter 3

Women's Changing Experiences of the Home and Life Inside it after Becoming Chronically Ill

Valorie A. Crooks

Introduction

Generally speaking, the home is a central part of everyday life; it is a space that is known for generating feelings or notions of privacy, sanctuary, and security (Imrie 2004, McDowell 1999). Houses are physical containers that become homes when we make sense of them using our 'personal meanings and associations' (Domosh and Seager 2001: xxii). Because we personally invest so much into making these places homes – through ascribing meanings, qualities, attributes, and associations – they, in fact, become complex spaces of everyday life. These spaces can also be quite complicated because the house itself may be the physical site in which multiple trajectories intersect, such as those of the lifeworlds of multiple family members who live together. Geographers' research on the home has highlighted the multitude of activities which take place within them, including those of: the giving and receiving of care by and for family members (e.g., Yantzi and Rosenberg 2008); the provision of formal home health care by paid workers (e.g., Lilly 2008); the decoration of the physical space (e.g., Domosh and Seager 2001); and the production of housekeeping work (e.g., McDowell 1999). The home also has a complicated relationship with people's health. Some of the characteristics associated with the home, such as sanctuary and security, can contribute to people's sense of wellbeing and thus how they think of and even experience health (Curtis 2004). Conversely, when people do not have a sense of home due to homelessness or the ways they feel about their house and the activities undertaken within, their health may be negatively affected.

In addition to the home being a potentially healthful space, in that its attributes may assist with gaining or maintaining a sense of wellbeing, it is also a highly gendered space. Central to this is what McDowell (1999: 79) refers to as the 'naturalized association of domestic work with women.' In the gendered division of labour women have traditionally taken responsibility for the bulk of the housework (e.g., laundry, shopping, and cooking) as well as care work (e.g., caring for children and ageing parents) undertaken within the home. As Pilcher and Whelehan (2004: 31) remind us: 'this domestic work/labour undertaken by women is unpaid, is mostly (though not always) performed within the home, and is necessary for the day-to-day maintenance of the household and its members.' Such work is thus a vital aspect of the home. In spite of this, it has also been, and continues to be, an undervalued contribution given the societal focus on capitalist production and its subsequent remuneration for paid workers (McDowell 1999). The undervaluing of work in the home even extends to paid work undertaken within. For example, Lilly (2008) has demonstrated that the work of health care aides in private homes is compensated for less than that which is undertaken in hospitals. At the same time, this gendered division of labour has resulted in particular spaces within the home becoming particularly feminized, especially those within which daily household work is undertaken. The kitchen, for example, is one such space and it is often built and/or laid out with women in mind (see Supski 2006, Saarikangas 2006). Because of factors such as the gendered division of labour within the home and also the feminization of specific home spaces we can imagine that women who are unable to do household work for any number of reasons or men who enjoy spending time cooking in the kitchen, among numerous other examples, challenge our notions of how the home should be read and understood.

The onset of chronic illness can result in the home taking on new meanings in people's lives (Angus et al. 2005, Dyck 2002). Dyck (2002), for example, contends that the meanings ascribed to one's 'home' can change as a result of women's restructuring of daily routines after the onset of chronic illness. Interpersonal relationships played out within the home are also subject to change after the onset of chronic illness. Söderberg et al. (2003) found that relationships between partners frequently become altered after women develop chronic illness. In their interviews with chronically ill women's husbands they found that outcomes for these men's home lives included: (1) an altered distribution of unpaid labour, in that they took on additional housework tasks, thus engaging in fewer recreational activities outside of the home as a result of added household responsibilities; and (2) the development of deeper relationships with children, due in part to spending more time on childrearing tasks. Öhman and Söderberg (2004) have also found that family members feel an increased sense of responsibility for chronically ill relatives and that they too must alter their daily lives in response to the person's illness which can lead to feelings of imprisonment within the home and a reduced sense of freedom. The literal spatiality of the home may also be experienced differently or in particular ways after the development of impairment or illness (Moss 1997). Imrie (2004) points out that the physical design and layout of the home can facilitate movement and mobility; something which is of great importance given that he contends that meaning is brought to people's everyday lives through how they experience the home. Thus, the onset of chronic illness can result in people experiencing a revised spatiality of, and life within, the home in multiple ways.

Fibromyalgia syndrome (FMS) is a chronic illness that primarily affects women (The Arthritis Society 2002, Clauw and Crawford 2003). The two main symptoms of FMS, chronic pain and fatigue (Clauw and Crofford 2003), are devastating and can result in bodily impairments which diminish women's abilities to fully participate in expected and/or desired ways in society and space (Åsbring 2001). Women's experiences of life with this chronic illness are the focus of this chapter. Specifically, the findings shared are drawn from a study involving interviews with women (n=55) after the onset of FMS living in Ontario, Canada. As discussed in greater detail elsewhere (Crooks 2007), the onset of FMS led to altered lifeworlds for all of the women involved in the study; where the lifeworld is comprised of 'the taken-for-granted mundane experiences of daily life as carried out in particular spatio-temporal settings' (Dyck 1995: 307). Developing symptoms such as mental haziness (popularly known as fibrofog) and sleep disorders left many with limited energy with which to engage in activities, an outcome of which was that 40 women had left paid labour due specifically to their health and most had altered their involvement in social and recreational activities. A result of such change was that places outside the home had become increasingly difficult for most of the women to access due to their restricted physical and mental abilities and, for many, diminished incomes as a result of no longer working. In other words, the home had taken on more significance and prominence in their everyday lives.

Herein I examine the ways in which the onset of FMS and continued negotiation of chronic illness alters women's uses of and relationships to and interpersonal relationships within the home. Drawing on the findings of the interviews mentioned above and also the results of a health-related non-condition-specific standardized quality of life instrument (the Sickness Impact Profile or SIP) administered to participants, I illustrate how these women's changing uses of and interpersonal relationships played out within the home have resulted, for many, in the development of an uneasy relationship with this space of everyday life. More specifically, in the remainder of this chapter I consider how the increasing presence of the home in the women's lifeworlds, their confusing and sometimes even contradictory understandings of the routines they live out within this site, and the altered interpersonal relationships they experience with those with whom they share this space collectively altered the women's relationships with the home as a site of everyday life. After sharing these findings I then move to introduce and discuss concepts that assist with contextualizing and explaining the women's reported experiences of this space, namely disablement, embodiment, and coping. I introduce and then specifically discuss these concepts as they relate to the empirical findings. In doing this I also reflect upon how some of changes reported by the women happened in particularly gendered ways; an important consideration given that the domestic space of the home is often thought of as womanly or feminine (England and Lawson 2005).

Study Background

Women who had developed FMS and were living in one of three communities in the province of Ontario, Canada – the cities of Hamilton, North Bay, and Sudbury – were sought out in this study. The study purpose was to document and explain the life-changing, and also simultaneous, processes of negotiating a changing socio-spatial life after developing a chronic illness and of becoming a chronically ill patient and negotiating the health care system. More specifically, the ways in which women's bodily experiences, roles and routines of everyday life, relationships with others, and negotiations of social institutions were affected by, and sometimes changed as a result of, their lives with FMS were examined. Women were recruited using a snowball sampling strategy and participated in semi-structured interviews which were analysed using the constant comparative technique. More detailed information about this part of the study can be found in Crooks (2006, 2007) and Crooks, Chouinard and Wilton (2008).

Every woman who participated in an in-depth interview also agreed to complete the SIP test. Designed by Bergner et al. (1981), the SIP was developed to test perceived health status using behavioural measures in samples of all sizes. It generates an overall quantitative measure of how significantly an illness impacts someone's everyday life using the concept of a continuum between health/function and illness/dysfunction (Damiano 1996). A more detailed description of the SIP analysis and its results can be found in Crooks (2007). Findings from this test are referenced in the analytic discussion that follows to support issues emergent from the interviews. More specifically, responses to individual questions that pertain to the focus of this analysis are included; the frequency of response is expressed both as a percentage and in 'n=x' format.

Interviews (including the SIP component) were conducted with a total of 55 women. Their ages ranged from 35 to 88 with the average being 58 years. On average they had lived with the symptoms of FMS for 14 years, with the shortest period being less than a year and the longest being 54 years. The average length of time since diagnosis was just under nine years (ranging from less than a year to 23 years). The women lived with an average of 1.3 chronic illnesses in addition to FMS, which ranged from a minimum of no other conditions to a maximum of five others. The women's dwelling types were quite varied. Twenty-seven owned family houses and lived with other family members, five widowed women owned family houses in which they lived alone, and five single women lived alone in owned houses. A further five women rented family houses, four rented subsidized family housing, and nine lived alone in rented housing.

An Altered Presence of the Home in the Women's Lifeworlds

Although the interviews were not intended to focus on the women's experiences of the home or home life, there was much discussion about their lives in this particular space after becoming chronically ill. This was certainly not surprising given the increasing amounts of time most of the women were spending at home after developing FMS. In this section of the chapter I draw on the interviews and SIP results to present three emergent themes related to how the women discussed home and life within it, namely: (1) their increased presence at home; (2) their unpredictably routine and routinely unpredictable home lives and activities; and (3) their interpersonal relationships at home.

An Increased Presence at Home

Due to the women's changing presences in paid labour and community and social life after the onset of FMS – in that 40 women had left the labour market due to health-related concerns, 12 were retired prior to onset or had never been involved in paid labour, and only 3 were employed at the time of the interview – the home had shifted from being *a* space of everyday life to *the* space of everyday life for all but a few participants. Sixty-nine per cent (n=38) of the women, for example, reported staying at home most of the time. And when activities were undertaken outside of the home, 69 per cent (n=38) also indicated that they stayed away from this place for only for brief periods of time in order to run errands such as grocery shopping. Further, trips to the mall, downtown core, and/or shopping areas had been eliminated as much as possible from daily life by 20 per cent (n=11) of the participants. Based on such findings it can be understood that these changes in the women's lifeworlds led to an increased presence of the home in their everyday lives after the onset of FMS.

An issue that underscored much of the women's discussion of the home was the frustration many experienced with regard to accomplishing less when at home than they did before becoming chronically ill while spending more time within this space than they did in their pre-FMS lives. Eighty-seven per cent (n=48), for example, indicated that they were doing less housework and fewer chores than before the onset of FMS. Sixty-seven per cent (n=37) said that they were not doing any of the normal maintenance around the house or yard that they once undertook, including heavy gardening and tasks such as painting, and 87 per cent (n=48) reported not doing any heavy work around the house. At the root of the women's reported frustration was that they were not able to accomplish things that they and sometimes even others expected them to do while at the same time not being able to engage in paid work. In fact, in their pre-FMS lives many of the women simultaneously undertook both paid and unpaid labour (e.g., housework, volunteer work). However, after developing FMS and departing from the workforce, the biggest 'contribution' most could make to the household was that of exclusively doing unpaid labour, when physically and mentally able to do so, such as laundry, cleaning, and meal preparation which is a reflection of the traditional gendered division of labour discussed above. Thus, traditional women's roles such as 'wife/ partner' and 'mother' loomed larger in their lives as others such as 'worker' were removed, while at the same time physical experiences of symptoms such as fatigue prevented them from doing some of the tasks ascribed to these roles.

Although the home had taken on a greater significance in the women's lives, this change did not occur without struggles over coming to terms with the loss of other parts of their socio-spatial lives that this shift signified. For example, the home had increasingly become a space within which rest, relaxation, and recuperation frequently took place in order to minimize FMS symptoms and enhance overall wellbeing. However, as noted above, many of the women struggled with spending increasing amounts of time at home. For example, Dee[1] was most relaxed when at home, but this physical comfort was uneasy in that, emotionally, she was less happy and sometimes even unsettled during periods of rest than in periods of productivity:

> I lost myself with fibromyalgia; I lost the person I am…And I don't really know the person I've become; it's foreign to me…I'm not a person that sits around. I'm not a person that doesn't do things. It's like I'm trapped…I don't understand this person…But when I'm at home and I can just lay in the chair I feel the most comfortable, it's when I'm in the least pain and you know if I'm exhausted I can just lay there, and, uh, but I'm not happy at that time, so I'm not me…The me is the person that's out there doing things and is happy, who's laughing and having fun, that's the me I know.

Dee's comment also reflects the reality that the increasing amounts of time spent in the home signalled other significant changes in the women's lives; changes such as dealing with accommodating chronic illness in their everyday roles and identities, mourning the loss of involvement in paid labour and/or being productive family members, and having reduced contact with friends in their lives. The home had thus become a place within which the women pondered their pre-FMS lives and existences, while also learning to come to terms with and negotiate, albeit uneasily, their less than fully able minds and bodies.

Activities at Home: Unpredictably Routine and Routinely Unpredictable

The women's everyday lives inside the home after developing FMS can best be described as simultaneously 'unpredictably routine' and 'routinely unpredictable'. The term 'unpredictably routine' characterizes those activities that made up the routine of home life, such as engaging in housework and gardening, that women undertook with temporal unpredictability or sometimes even stopped doing

---

1   In order to maintain anonymity pseudonyms chosen by the participants are referred to.

completely after becoming chronically ill. Life was simultaneously reported to be 'routinely unpredictable' in the home in that activities, whether routine such as vacuuming or not such as hosting a dinner party, were frequently temporally and even spatially modified, interrupted, discontinued, or cancelled as a result of the bodily realities with which they were living on any particular day. This finding is significant as it tells us that how the women read and understood their home lives was connected to the troubling unpredictability of their FMS symptoms, thus rendering life in this space, confusingly, both routinely unpredictable and unpredictably routine.

A result of the situation described above was that the women often discussed having some semblance of a routine at home while also talking at some point about the unsettling unpredictability not only of their bodies but also of their lives in terms of what they were or were not able to accomplish at a given point in time. Or, oppositely, some women clearly stated that they had no routine but were then able to discuss clearly routine elements of their home lives. Sheila and Moe's comments reflect this situation:

> Sheila: I don't have a daily routine, um, because everyday is different. It's like putting your hand in a box of bits and bites…like every day is different… Depending on what time of the year it is, the kids go to school or not, you…get your kids off to school and work or whatever, and uh, then I'll go to bed. And lay down for two or three hours or whatever. And then, what do I do? Maybe I can do something, maybe I can't. I don't do any major cleaning. I can do laundry…Have one day a week I make myself get out for lunch with a girlfriend, her and I meet for lunch, and um, but that's all we can do is go for lunch, we can't go shopping or anything.

> Moe: I don't think I have a routine…I sort of do what I have to do, or you know, what I feel like doing. Well, usually first thing I eat or walk the dog, first thing. Or I do my stretching exercises first thing. But I always, just about ninety-nine per cent of the time I do about half an hour of stretching exercises in the morning to get me going. And I've…for the past year I've been going to tai chi classes two or three times a week…My husband and I are both retired, so it's do what we feel like doing, sort of thing.

Although characterized as non-routine by Moe and Sheila, these descriptions of daily life both within and outside of the home certainly sketch out the structure of a spatio-temporal routine that was far from completely chaotic or unpredictable despite their discussions of living a (home) life lacking in routine.

Although most of the women's days were centred on the home space, their abilities to undertake tasks within this space were diminished, as noted previously. One strategy employed to assist with accomplishing household tasks was to pace activities. For example, ninety-six per cent (n=53) of the women indicated that they did work around the house for only short periods of time and took frequent rests and breaks. The inclusion of breaks within the day was another element of the women's unpredictably routine yet routinely unpredictable home lives, in that rest breaks and naps at home had become part of everyday life for most of the women but were not taken in a temporally or even spatially routine manner. For example, the women frequently reported taking breaks whenever and wherever needed in the house, particularly when undertaking energy-consuming tasks such as cleaning, cooking, and vacuuming, but also talked about the need for such breaks as being a reflection of their unpredictable lives. A mid-afternoon nap on the couch could take place one day while the on next day a rest may have included sitting at the kitchen table for 20 minutes during meal preparation. The bedroom had become a more central sphere of home life for many of the women in this regard in that it had become a site for rest breaks. More specifically, 49 per cent reported (n=27) staying in bed more during the waking hours.

Experiencing Interpersonal Relationships at Home

The home was experienced as an isolating space for many of the women, 67 per cent (n=37) of whom reported spending much of their time alone within it. Some of this isolation was desired, in that 22 per cent (n=12) indicated that they actively isolated themselves from family and 13 per cent (n=7) reported refusing contact with and turning away from family, while much of it was the undesirable outcome of being at home when others were at school, at work, or were engaging in social lives of their own.

Interactions and relationships with others who shared the home changed significantly as a result of developing FMS. Seventeen of the women talked about the changed or changing nature of their relationships with partners or spouses and thirteen commented specifically about their spouse or partner taking on additional tasks around the house that they could no longer effectively do. Those with partners or spouses involved in paid labour outside the home were particularly cognisant of the strain this put on their relationships and often felt guilty for placing added demands on their home time. Gloria's comment conveys this guilt:

> You know, if I'm not up to it [housework], I don't do it. If I'm tired I lay down. I can't do some things. Like, vacuuming is one of the hardest things to do. I'll ask my husband 'Could you please do it?' The one time I asked him I said 'Do you find it hard vacuuming? Is it hard for you?' He said 'no.' I said 'Oh good!' I said 'I'll ask you to do it all the time, then!'…But just there's that feeling of guilt asking him because he works twelve hours…and because I am not working, it's really…I have that guilt that I should be doing…all this stuff at home. But, you know, you try to let that go.

Agnes' husband found the increasing expectations being placed on him when at home to be difficult to negotiate: 'whereas I have the opportunity to lie down sometimes he says "Well you know, you just sit all day, you know, if you don't feel good and if I'm tired I still have to go to work. So it's not fair you expect all this of me.'" And Courtney told of how FMS 'sure doesn't do much for a marriage.' Another three women talked of how their spouse or partner's lack of understanding of their FMS created strain in the relationship and two discussed acting moodier in interactions than they did in their pre-FMS lives. This is not surprising given that 24 per cent (n=13) reported acting increasingly disagreeable, in terms of being spiteful or stubborn, with family members.

In terms of changed or changing relationships with children, thirty women talked about this during the interviews. For nine women, delegating tasks and responsibilities within the home to their children had become a strategy adopted in order to better manage getting chores completed. Seven women reported feeling as though their children did not fully understand the nature of FMS, which led to frequent misunderstandings and sometimes arguments and even long-standing anger or resentment. Irene B. said this:

> 'They [her children] just, they didn't know what in God's name was wrong with me. They were like, you know, like "Mom, go to work." And I'm like "I can't." Like I just found the relationships really suffered.' Ann's relationship with her daughter also suffered: 'My daughter is very angry with my illness, which is translated into being very angry with me for a long time.'

Such changes in interpersonal relationships affected how interactions played out within the home.

As a result of limited energy and impaired bodies, eight women talked about engaging in fewer activities with their children such as taking them to sporting events or other sites of entertainment outside of the home. The women frequently expressed feelings of guilt over having done so; as Courtney told: 'When this [the onset of FMS] happened my daughter was only two years old. Now she's twenty-seven, [and has] got two kids of her own…I felt like I was cheating her out of a real mother.' Another nine women talked about limiting activities with their grandchildren, both within the home and beyond. Although many women prioritized interactions with their children, they perceived these relationships to have been negatively affected, if not permanently altered, because of a decreased involvement in their children's lives and increased responsibilities they had placed on them both when they were living at home (e.g., household chores) and sometimes even when they were not.

As per the findings shared above, the development of FMS not only resulted in changes to the women's own lives and lifeworlds but also those of their immediate family members with whom they shared the home space. Such changes left the women not only mourning for their pre-FMS selves and bodies but also with feelings of guilt because responsibilities they considered to be their own were being shouldered by others. Their lifeworlds also shrunk in gendered ways not only in that they spent more time in the domestic and 'womanly' space of the home but also because those women with (grand)children, and especially ones living in the family home, prioritized (grand)mothering over other activities that were once part of their lives such as visiting with friends. Despite this, 47 per cent (n=26) of the women felt as though they were not able to do the things they normally would to take care of their children and families. It can be understood that internalized notions of what it means to be a 'mother' and/or 'wife/partner' and ideas about to how to continue making a contribution to the family unit despite no longer undertaking paid work are some of the social forces that shaped how the women experienced life at home after the onset of FMS.

Explaining the Women's Experiences of the Home

Evidenced in the previous section, the women's negotiations of life at home and their relationships with and within this particular place after the onset of FMS were complex, at times challenging even their most basic senses of self (e.g., as 'productive' individuals). Further, the ways in which they ascribed meaning(s) to their homes had also changed alongside their uses of this space. For example, some of the women's feelings of unsettlement while spending increasing amounts of time in this space challenge popularized notions of the home as a sanctuary and restful place. Related to this, some of the women intertwined their understandings of their homes and their bodies when expressing that unpredictable minds/bodies can lead to unpredictable home lives. How do we make sense of these and other findings shared in the previous section? There are many potential ways this can be done, one of which is to put them into context. In this section of the chapter I contextualize and thus further unpack the findings by drawing on three particularly useful concepts: (1) disablement, (2) embodiment, and (3) coping. These concepts help to build an understanding of *why* the women experienced home in the ways they did after becoming chronically ill.

Disablement

Disablement is something that may be experienced as a result of becoming chronically ill. The findings shared in the previous section regarding the women's negotiations of the home after developing FMS show that some of them were experiencing this space in disabling ways which, in turn, affected how they both used and understood their homes. Understanding the process of disablement involves conceptualizing the interconnections between the mind, the body, and lived realities, including decisions and choices made that mediate the interaction between the physical body and socio-spatial life. According to Verbrugge and Jette (1994: 1), the disablement process '(1) describes how chronic and acute conditions affect functioning in specific body systems, generic physical and mental actions, and activities of daily life, and (2) describes the personal and environmental factors that speed or slow disablement, namely, risk factors, interventions, and exacerbators.' Crooks and Chouinard (2006: 346) conceptualize it to be 'an embodied process of becoming and being ill in place(s) and over time in ways that have disabling outcomes for individuals.' Further, Driedger, Crooks and Bennett (2004: 121) contend that this process leads to 'spatial deprivation and socio-economic exclusion as a result of the outcomes of the disablement process happening over both space and time and also in place.'

Importantly, disablement involves factors that are both within and beyond a person's control (Verbrugge and Jette 1994). For example, after the onset of FMS a woman may seek workplace accommodations that would allow her to maintain involvement in paid labour, this being a choice that is within an individual's control. Such a decision is constrained, however, by whether or not her employer is willing and/or able to make such changes and also her own ability to identify modifications that would appropriately accommodate her fluctuating ability levels. This same woman may also be negatively affected by her employer's ability to finance such accommodations, which is the result of institutional and administrative systems that are beyond her immediate control. These factors together, both those within and beyond her individual control, shape and constrain her ability to maintain involvement in paid labour and can ultimately inform her increased life within the home. Although being used as an example here, it was under these very circumstances that some of the women interviewed in this study ended up leaving paid work and spending more time at home.

The concept of disablement assists with putting into context some of the findings shared in the previous section. With regard to the women's increased presence in the home over time after the onset of FMS, the SIP findings confirmed that there were many factors that led to this outcome. Some of these factors and their outcomes were indeed within the women's control (e.g., (often constrained) decision-making regarding leaving paid work) while others were not (e.g., physical and financial access to spaces outside the home). Outcomes of experiencing the process of disablement over space and time were also numerous for the women, including experiencing social and spatial isolation in their homes and also changed relationships with those they shared this space with. Further, many of the women were struggling with some of the outcomes of engaging in the process of disablement after becoming chronically ill, and specifically with spending more time at home. Because of this the home had become a space of unease at times. This is perhaps best exemplified by Dee's comment: 'when I'm at home and I can just lay in the chair I feel the most comfortable, it's when I'm in the least pain...but I'm not happy at that time.' This sense of uneasiness, as demonstrated here through feeling physically comfortable but emotionally uncomfortable, can clearly have a negative impact upon the ways in which the home may contribute to the women's overall wellbeing or even specifically to their health; this being another outcome of the process of disablement as it relates to the home.

Embodiment

Alterations to the lifeworld made over space and through time, such as throughout experiencing disablement, are intimately connected to women's ill bodies and the ways in which illness is embodied. This was certainly true for how the women experienced the home after the onset of FMS. Most basically, as Charmaz (1995) suggests, changes in the body lead to changes in chronically ill women's lives. This is because 'the physical body and the social placings are viewed as intertwined... [each] affecting experiences of health, ability, impairment, disability and chronic illness' (Driedger, Crooks and Bennett 2004: 124). While there may be social, economic, spatial, and other types of outcomes resulting from the development of a chronic illness, including that of experiencing disablement as discussed above, the fact that these changes are rooted *both* in a woman's ill body and in larger social and institutional systems cannot be denied (Hall 2000, Moss and Dyck 2002).

Embodied expressions of chronically ill womanhood are central to an understanding of how women live out their altered lives after the onset of illness (Crooks and Chouinard 2006), including within the home. Embodiment, defined here as ways of expressing particular experiences, roles, identities, and feelings through the body, is highly relevant to the focus herein. This is because this concept draws attention to the different ways in which chronically ill people deal with and give expression to bodily experiences, such as pain, and the bodily expectations associated with social roles and identities, such as friend, including highly gendered ones, such as mother or wife. It is not uncommon, for example, for people with invisible or hidden chronic illnesses such as FMS to 'work through' their bodily pain in order to visibly maintain identities or roles which are socially constructed as being normal when in the presence of others (Rosenfeld and Faircloth 2004). In such an instance they attempt to embody an identity or role in a way that makes them appear as normal in spite of bodily realities such as pain and fatigue. Given this, the findings indicate that the home can be a site of undertaking acts of accepting or rejecting ill womanhood in embodied ways.

The concept of embodiment is useful for contextualizing some of what the women reported about their experiences of the home and interpersonal relationships experienced within it after becoming chronically ill. Clearly, the women lived out their lives in their homes in bodily ways. Bodily pain and fatigue very much regulated the activities they undertook both at home and beyond. The SIP findings also point to the fact that the women avoided certain kinds of bodily work at home after developing FMS, such as heavy lifting and housework. The women's discussions of their home lives and routines also raised issues of the body, and specifically a relationship, if not conflation, between bodily unpredictability and unpredictability in routinely completing household tasks and other activities. Despite these bodily realities, most of the women did their best to embody the roles of spouse/partner, mother, and household member, including in gendered ways and especially when others were around, out of a desire to appear productive in particular and also to maintain valued relationships with others.

Coping

Coping strategies are often developed after the onset of chronic illness as a way to deal with both physical and mental stress and stressors. In the broadest sense, the two main types of coping are: problem-focused and emotion-focused (DeCoster and Cummings 2004, Stetz, Lewis and Primomo 1986). Problem-focused coping takes place when one identifies and adopts a strategy that will assist with engaging in activities of daily living as a way of minimizing the impact associated with change or adjustment. This may include the insertion of rest periods during arduous tasks in order to better manage pain and lessen mental fatigue, which was something the women reported doing in order to manage household tasks. Emotion-focused coping centres on managing one's inner states, and thus developing ways to come to terms with one's altered life, sense of self, and place in the world. Placing focus on what one is able to do instead of what one can no longer do is an example of such a coping strategy.

The two types of coping strategies mentioned above are not mutually exclusive in that both can have emotional outcomes and/or centre on particular altered life circumstances or situations as they relate to chronically ill women's experiences of the home. For example, as Bondi, Davidson and Smith (2005) suggest, emotions are transformed as individuals experience things that change and/or alter their lives. Thus, the very nature of learning to live with and embodying chronic illness while experiencing disablement and developing coping strategies that alter life in-place, including within the home, is an emotional experience. At the same time, because we have an emotional involvement with places such as our homes (Bondi, Davidson and Smith 2005), when we alter how we live out our lives within them as a way of coping with change then our emotional relationship to them will also be altered. Thus, changing how one lives everyday life within the home may also alter how one experiences feeling (emotionally) connected to this space after the onset of chronic illness through the process of coping. This has clear implications both for how meanings are assigned to the home and also for the ways in which the home may contribute to a sense of wellbeing for chronically ill women.

The findings shared in the previous section have much to do with coping and the specific strategies to enact it that the women had adopted. More specifically, coping with pain and fatigue necessitated both emotion-focused and problem-focused strategies for their lives at home. Both types of coping strategy also necessitated making some things about home life routine (e.g., taking rest breaks or naps) while at the same time breaking with other routines (e.g., interrupting regular parts of home life to accommodate a rest or nap) in order to allow activities to be undertaken. The findings revealed that the women had to cope emotionally with their feelings associated with the loss of productivity and sometimes even inabilities to meet gendered norms and expectations regarding the enactment of roles such as mother, grandmother, and/or partner/spouse. Examples of how this took place included re-valuing that which they were able to do and also giving new meanings to terms they may use to describe themselves such as 'ill' or 'disabled' (see Crooks, Chouinard and Wilton 2008). Such emotion-focused coping strategies assisted with making their lives at home more tolerable. Taking rest breaks during tasks and pacing activities, for example, assisted with coping with FMS symptoms and was an important problem-focused strategy. At the same time, some of the problem-focused coping strategies they employed, such as not taking too many trips for shopping or going to town, in fact relegated them to the home, thus adding to the complexity of the relationship between coping and the home. Thus, staying at home more was a coping strategy in and of itself while at the same time the increased presence of the home in their lives was something to be coped with.

Conclusion

In this chapter I have drawn on the findings of interviews and a health-related standardized test to explore women's altered experiences of the home and life inside it after the onset of chronic illness. Three themes emerged from the dataset about this issue: (1) the women's increased presence at home; (2) the women's unpredictably routine and routinely unpredictable home lives and activities; and (3) the women's interpersonal relationships at home. Three concepts were then introduced which assist with contextualizing and explaining the findings reported: (1) disablement; (2) embodiment; and (3) coping. These concepts aid in understanding *why* women reported the experiences of the home and relationships within it that they did. The findings also revealed that the women struggled in particularly gendered ways, both implicitly and explicitly, with renegotiating their new presences in the home. For example, the women tried as much as possible to continue to do 'traditional women's work' within the home, including embodying the productive social roles they valued despite bodily pain and fatigue and the negative outcomes of experiencing disablement, prioritizing it over social and recreational activities, and developed coping strategies to assist them completing role-related tasks.

The interviews revealed that while the women's involvement in certain spaces of their pre-FMS lives had diminished or been eliminated completely, such as participation in paid labour at work or volunteer labour in the community (see also Crooks 2007), the home and activities undertaken within it loomed larger in their lifeworlds. Their emotional relationships to the home had been altered in that its increased presence in their lifeworlds signalled that other roles, such as that of paid worker, and spaces, such as those within which recreational activities had been pursued, that held significance in their lives were no longer part of everyday life. This, in turn, led to the women ascribing new meanings to the home, some of which deviated from socially expected norms such as understanding one's house to be a sanctuary or even health promoting. The SIP findings illustrated how, for the most part, the women's social lives within and beyond the home had diminished and their social and familial relationships had been altered as after the onset of FMS and some of the problem-focused coping strategies they had adopted. Such changes left the women struggling to come to terms with what it meant to be a mother, spouse/partner, woman, and/or productive family member with FMS in the home, often in emotional ways.

As Verbrugge and Jette (1994) remind us, both personal factors and environmental factors can speed up or slow down how someone experiences the process of disablement after the onset of chronic illness. For these women, the departure from paid labour and centring of the lifeworld around the home was a significant benchmark in how they experienced this process in that it signalled, for most, the point at which their embodiment of the chronically ill woman role and experiences of FMS prevented them from taking part in roles they had once considered central to their lives and identities.

Finally, there was a changed spatiality to the women's lives within the home. For example, despite the centring of the women's lifeworlds around this space, activities engaged in within it had taken on new meanings and were not undertaken in the same fashion or capacity as in their pre-FMS lives. Despite a majority of the women reporting that they spent most of their time in the home and discussing activities that shaped a temporal structure to their daily lives, many described their lives as lacking in routine. This led to 'unpredictable routine' and 'routine unpredictability' in their daily spatio-temporal lives within the home. More specifically, the findings illustrate that the women's daily lives were relatively spatially routine ('unpredictably routine') yet simultaneously temporally dynamic ('routinely unpredictable'). One reason for this is that many of the women conflated fluctuating and unpredictable bodily symptoms with having unpredictable and non-routine everyday lives. Such a finding assists in unpacking the notion of 'daily routine' in chronically ill women's lives by differentiating between spatial routine and temporal routine, including at home specifically. It also reflects a closeness between women's understanding of their bodies and their homes. Life in the home was also spatially isolating in that the women's abilities to maintain a social life were compromised by their inabilities to engage in activities with others both inside and outside the house in predicable ways. All of these findings point to the fact that we must carefully consider the complex connections between factors such as home and health, home and body, and home and gendered roles if we are to enhance our knowledge of how women negotiate life with chronic illness.

Acknowledgements

First, I must thank all the participants for sharing their stories. This research was supported by a grant from the Arthritis Health Professions Association. Stipendiary funding was provided by a Social Sciences and Humanities Research Council of Canada doctoral fellowship, a Centre for Health Economics and Policy Analysis doctoral studentship, an Ontario Graduate Scholarship, and the E.B. Ryan Scholarship for Research on Aging.

References

Angus, J., Kontos, P., Dyck, I., McKeever, P. and Poland, B. 2005. The personal significance of home: habitus and the experience of receiving long-term care. *Sociology of Health and Illness*, 27(2), 161-187.

Åsbring, P. 2001. Chronic illness – a disruption in life: identity-transformation among women with chronic fatigue syndrome and fibromyalgia. *Journal of Advanced Nursing*, 34(3), 312-319.

Bergner, M., Bobbitt, R.A., Carter, W.B. and Gibson, B.S. 1981. The sickness impact profile: development and final revision of a health status measure. *Medical Care*, 19(8), 787-806.

Bondi, L., Davidson, J. and Smith, M. 2005. Introduction: geography's 'emotional turn', in *Emotional Geographies*, edited by J. Davidson, L. Bondi and M. Smith. Burlington and Aldershot: Ashgate, 1-16.

Charmaz, K. 1995. The body, identity, and self: adapting to impairment. *The Sociological Quarterly*, 36(4), 701-724.

Clauw, D.J. and Crofford, L.J. 2003. Chronic widespread pain and fibromyalgia: what we know, and what we need to know. *Best Practice and Research in Clinical Rheumatology*, 17(4), 685-701.

Crooks, V.A. 2006. 'I go on the Internet; I always, you know, check to see what's new': chronically ill women's use of online health information to shape and inform doctor-patient interactions in the space of care provision. *ACME: An International E-Journal for Critical Geographies*, 5(1), 50-69.

Crooks, V.A. 2007. Exploring the altered daily geographies and lifeworlds of women living with fibromyalgia syndrome: a mixed-method approach. *Social Science and Medicine*, 64(3), 577-588.

Crooks, V.A. and Chouinard, V. 2006. An embodied geography of disablement: chronically ill women's struggles for enabling places in spaces of health care and daily life. *Health and Place*, 12(3), 345-352.

Crooks, V.A., Chouinard, V., and Wilton, R.D. 2008. Understanding, embracing, rejecting: women's negotiations of disability constructions and categorizations after becoming chronically ill. *Social Science and Medicine*, 67, 1837-1846.

Curtis, S. 2004. *Health and Inequality*. London: Sage.

Damiano, A.M. 1996. *SIP: Sickness Impact Profile: User's Manual And Interpretation Guide*. Baltimore, MD: The Johns Hopkins University, Medical Outcomes Trust and Health Technology Associates Inc.

DeCoster, V.A. and Cummings, S. 2004. Coping with Type 2 diabetes: do race and gender matter? *Social Work in Health Care*, 40(2), 37-53.

Domosh, M. and Seager, J. 2001. *Putting Women in Place: Feminist Geographers Making Sense of the World*. New York: The Guilford Press.

Driedger, S.M., Crooks. V.A. and Bennett, D. 2004. Engaging in the disablement process over space and time: narratives of persons with multiple sclerosis in Ottawa, Canada. *The Canadian Geographer*, 48(2), 119-136.

Dyck, I. 1995. Hidden geographies: the changing lifeworlds of women with multiple sclerosis. *Social Science and Medicine*, 40(3), 307-320.

Dyck, I. 2002. Beyond the clinic: restructuring the environment in chronic illness experience. *OTJR: Occupation, Participation, and Health*, 22, 52-60.

England, K. and Lawson, V. 2005. Feminist analyses of work, in *A Companion to Feminist Geography*, edited by L. Nelson and J. Seager. Oxford, UK: Blackwell Publishing, 77-92.

Hall, E. 2000. 'Blood, brain and bones': taking the body seriously in the geography of health and impairment. *Area*, 32(1), 21-29.

Imrie, R. 2004. Disability, embodiment and the meaning of the home. *Housing Studies*, 19(5), 745-763.

Lilly, M.B. 2008. Medical versus social work-places: constructing and compensating the personal support worker across health care settings in Ontario, Canada. *Gender, Place and Culture*, 15(3), 285-299.

McDowell, L. 1999. *Gender, Identity and Place: Understanding Feminist Geographies*. Minneapolis: University of Minnesota Press.

Moss, P. 1997. Negotiating spaces in home environments: older women living with Arthritis. *Social Science and Medicine*, 45(1), 26-33.

Moss, P. and Dyck, I. 2002. *Women, Body, Illness: Space and Identity in the Everyday Lives of Women with Chronic Illness*. Walden, MA: Rowman and Littlefield Publishers.

Öhman, M. and Söderberg, S. 2004. The experience of close relatives living with a person with serious chronic illness. *Qualitative Health Research*, 14(3), 396-410.

Pilcher, J. and Whelehan, I. 2004. *50 Key Concepts in Gender Studies*. Thousand Oaks, CA: Sage Publications.

Rosenfeld, D. and Faircloth, C. 2004. Embodied fluidity and the commitment to movement: constructing the moral self through arthritis narratives. *Symbolic Interaction*, 27(4), 507-529.

Saarikangas, K. 2006. Displays of the Everyday: Relations between gender and the visibility of domestic work in modern Finnish kitchens from the 1930s to the 1950s. *Gender, Place and Culture*, 13(2), 161-172.

Söderberg, S., Strand, M., Haapala, M. and Lundman, B. 2003. Living with a woman with fibromyalgia from the perspective of the husband. *Journal of Advanced Nursing*, 42(2), 143-150.

Stetz, K., Lewis, F.M. and Primomo, J. 1986. Family coping strategies and chronic illness in the mother. *Family Relations*, 35(4), 515-522.

Supski, S. 2006. 'It Was Another Skin': the kitchen as *home* for Australian post-war immigrant women. *Gender, Place and Culture*, 13(2), 133-141.

The Arthritis Society. 2002. *Fibromyalgia*. The Arthritis Society of Canada. Available online at: www.arthritis.ca/types [accessed 19 September 2003].

Verbrugge, L. and Jette, A. 1994. The disablement process. *Social Science and Medicine*, 38(1), 1-14.

Yantzi, N.M. and Rosenberg, M.W. 2008. The home as a place of care: the contested meanings of home for women caring for children with long-term care needs. *Gender, Place and Culture*, 15(3), 301-315.

Chapter 4

Enabling Cultures of Dis/order Online

Joyce Davidson and Hester Parr

Introduction

This chapter considers how virtual space helps constitute particular geographies of difference for those on the Autistic Spectrum (AS)[1] and those with general mental health problems such as Anxiety Disorders (ADs) through an analysis of online writing – and off-line writing about online experience – amongst these groups. Neither of these 'groups' (diversely composed of a broad spectrum of individuals, experiences, cultures and 'conditions' (Davidson 2008b, Parr 2008)) would *necessarily* identify as 'disabled'. And yet, one key reason why they inhabit online spaces is precisely because off-line environments can feel sensorially overwhelming, emotionally uncomfortable, socially stigmatizing and so, *disabling*, rendering those with AS and mental health difficulties vulnerable and excluded in many different ways. Previous research in this area has suggested that virtual space offers different kinds of potentially 'therapeutic geographies' (Davidson and Parr 2008), where new kinds of trust (Parr and Davidson 2008) are established between people otherwise unconnected in various recovery and support scenarios. In what follows we wish to explicitly compare the online experience of those on the AS with those using mental health forums for virtual exchange and support.

Direct comparisons between such very different groups might be considered unusual and perhaps even inappropriate, and there are certainly important distinctions between the two as detailed in this chapter.[2] Geographers are, by

---

1   AS encompasses a wide array of neurodevelopmental conditions typically characterized by challenges in social interactions and communication skills, and the presence of repetitive behaviours and restricted interests (APA 2004).

2   There are two separate studies that contributed empirical materials to this case study. The first, focused on the Internet and mental health problems, is funded by the UK Economic and Social Research Council (RES-000-27-0043). In this study an extensive online Internet user survey designed with a combination of both closed and open-ended questions was completed by 78 respondents. The survey was advertised on four UK mental health discussion forums, three of which were explicitly user-led, and these included 'Mental Health in the UK' (www.zoo.pwp.blueyonder.co.uk: MHUK), 'The Mental Health Foundation' (www.mentalhealth.org.uk: MHF), 'Little Wing' (www. littlewing.org.uk: LW) and 'The National Phobics Society' (www.phobics-society.org.uk: NPS). Quotations from survey responses are used in anonymized form. The survey in

now, relatively well acquainted with life-worlds disrupted by anxiety, depression, compulsions and other conditions of mental and emotional ill-health (Parr 2008, Segrott and Doel 2004). Autism is very different, even if its parameters are similarly authoritatively defined in the American Psychiatric Association's (2004) manual of *mental disorders*. Studies of autism have been far less frequent, and the condition far less familiar to health geographers; with published explorations of first-hand accounts of the spectrum only recently appearing (Davidson 2007, 2008a, 2008b). These personal accounts provide insights into social and emotional life with AS of a kind largely absent from far more common parental (e.g., Grinker 2008, Iverson 2008) and professional (e.g., Aarons and Gittens 1999, Szatmari 2004) versions of these 'disorders of affect'.[3] Such second-hand accounts emphasize affective challenges around social interactions and communication skills and suggest those with autism are largely un-emotional and a-social, presenting repetitive

---

the study was followed up by telephone interviews with five respondents and moderators of each web-based forum. All materials were coded and analysed. These methods were supplemented with online overt ethnography in each of the sites, including participation in various discussion threads and live chat-rooms. The second study is part of a project funded by Social Sciences and Humanities Research Council of Canada. The primary source material for this chapter is comprised of published autobiographies by those who claim a place on the AS. Forty-five autobiographies and edited collections of first-hand accounts were identified through various literature searches, and with the research assistance of Leah Huff and Victoria Henderson. These texts were subjected to a sequential process of detailed annotation, coding and critical discourse analysis (Fairclough 1995, Smith 1996) before data-saturation was judged to have occurred. Autobiographical texts have been utilized in very similar ways to other forms of primary source materials – e.g., transcripts from interviews conducted during past research with sufferers from phobic and anxiety disorders (Davidson 2003). It should, however, be acknowledged that many of those on the spectrum lack the skills or resources, whether cognitive, social or financial, to present their accounts in a publishable form. Only the most coherent accounts are likely to be published and such accounts will usually be written by those at the less 'severe' end of the spectrum, who tend also to have access to, and ability to navigate, virtual social space. Many AS individuals excluded from this study will no doubt have different, less enabling experiences of the Internet, and alternative research design and methodologies would be required to gain access to such experience.

3   The term AS is used here to reflect the preference of the majority of authors whose work contributes to this study. Many find the inclusive connotations of this diagnostic term preferable to other more contentious and potentially divisive labels in clinical circulation, such as 'low functioning' or classic Kanner's autism, or supposedly higher-functioning Asperger's Syndrome (Shore 2001). AS is seen to facilitate constructive focus on experiential commonalities across the spectrum in a manner faithful to the aims of advocacy many support. In this paper, the term 'autism' is at times used as shorthand, but is still intended to denote the full spectrum. However, we do not wish to suggest these terms are accepted by all. For recent discussions of self-definitions of autism, see Bagatell 2007, Baker 2006, Brownlow and O'Dell 2006, Clarke and van Amerom 2007, Gevers 1993, Jones et al. 2001, Waltz 2005.

behaviours and restricted interests of an atypically *solitary* or 'self-centred' kind (the diagnostic label in fact derives from the Greek *autos*, meaning 'self', see Stanghellini 2001). Personal accounts of social life on the spectrum often suggest otherwise, revealing profound interest in, and feeling for, others (Biklen 2005, Davidson 2007), and while AS emotional geographies are certainly complexly, a-typically constituted, emotionality and sociability are not always or entirely absent; just because being with others proves difficult does not mean it is not desired, as evidenced by experience of many and varied mental health problems. Despite, then, many possible differences between those with affective (autistic) and emotional (mental) disorders, what *connects* members of these groups is a form of socio-spatial exclusion, imposed or self-sought, because of challenges with real time socio-spatial interaction, and an associated need to find alternative spaces to connect with (similar *and* different) others.

The analysis below suggests that a broad range of outcomes result from Internet use between and among members of these groups. There are many striking similarities yet significant divergence in experiences, and these various outcomes have implications for considerations of difference and sameness (or 'deviance' and 'normality', see Moser 2006). In turn, these considerations help us understand a little more about the Internet as a technology which disassembles, enables or perpetuates disabling geographies. These arguments are explored below through an explicitly spatial language, whereby the Internet is understood as a space in which powerful social relations are productive of particular cultures, which in turn hold the potential to render 'different' groups more or less proximate to those (falsely) considered 'mainstream' or 'typical', un-problematically 'able-bodied' and 'able-minded'. This is a story, then, about distance and proximity and the (re)construction of geographies of difference and sameness.

To elaborate these themes a little further, let us first turn to studies that have considered technologies and relations of ableism (Gleeson 1999). To think of the Internet as a particular technology important in dismantling geographies of difference for vulnerable and stigmatized groupings with mental health issues and AS characteristics is to position this debate in a long line of potentially emancipatory 'solutions' to disabled difference (Hahn 1989, Moser 2006). In reference to off-line technologies designed to 'overcome' the barriers of the built environment for disabled people, Gleeson (1999: 99) points to a range of studies that contain 'rather uncritical faith in the power of technologies to overcome the "limitations of disabilities"'. Whilst acknowledging that technical aids can be useful to some people, Gleeson argues that the risks of assuming technological fixes to the 'disabled problem' can risk rendering any discussion of disability as only a 'natural', rather than also as a social, state. Moser (2006) agrees, as she argues that technologies deployed in these circumstances often work to build particular *orders* of normality and deviance through which the abled and disabled are understood. Although technologies may assist mobilities (usually physical ones) for example; they usually operate within a 'normalising and compensatory logic' (Moser 2006: 388) that ultimately works to reproduce relations of difference that often constitute experiences of disability (i.e., the disabled person is only 'made normal' through the technology). Such an argument is particularly important when considering the lives of physically disabled people. However, we wonder whether the same critique can be advanced in relation to those with 'mental', emotional or neurological differences (see Davidson 2008c), and for differences that are either irreversibly organic or recoverable? In what follows then, we draw on primary source materials from our different research projects – online survey responses and autobiographical texts[4] – to assess experience of the Internet as a technology that may or may not have emancipatory implications for these groups. We specifically take up the challenge of thinking about how the Internet may work to produce orders of normality and difference for people living with AS and mental health problems in both on and off-line worlds.

Internet Geographies and Cultures of Difference

Much has been written about the emancipatory potential of the Internet in terms of general embodiment and health (for references, see Parr 2002, Crooks 2006, Madge and O'Connor 2006) and the benefits of online participation for people with mental heath problems in particular (Davidson and Parr 2008, Parr 2008, Wooten, Yellowlees and McClaren 2003). Virtual geographies of communality between people with mental health problems are arguably one innovative solution to what some might call a *crisis* of 'real-space' community and its caring capacities for others' difference. In recent years the Internet has become an important geography of support, therapy, leisure, self-work and advocacy for both individuals with autism and those with mental health problems, at least, that is, for those with sufficient material and/or social resources to facilitate access.[5] Thinking about the Internet as a technology that might enable a proximate recognition of difference(s) in their own terms, rather than exist *just* or merely in a 'compensatory logic' (Moser 2006) is particularly important when thinking about AS, as a form of neurological difference (Davidson 2008c) and given the arguments above. Internet writings can stress the iterative and rapid competencies of autistic individuals in ways previously unrecognized by others. While bearing in mind questions of representation

---

4   The specific methodological implications of using autobiographies as primary source materials are discussed at length in several recent texts, for example Smith (1996), Smith and Watson (2001) and Avrahami (2007). Clearly, self-authored life narratives provide an invaluable yet under-explored qualitative resource for those interested in understanding 'insider' accounts of different or disordered experience, such as that associated with AS.

5   According to Eng et al. (1998: 1371), additional 'barriers to access include cost, geographic location, illiteracy, disability, and factors related to the capacity of people to use these technologies appropriately and effectively'. Internet access is not – nor ever will be – universal. Such limitations should be borne in mind when drawing conclusions about this technology's enabling potential (see Parr 2008: 136).

(see footnote 2), popular periodicals and mainstream media nevertheless report regularly on the conspicuous autistic presence online. For example, a recent *New Scientist* article (Biever 2007: npn), states that 'Since the 1990s, people with autism have been communicating via chatrooms, email lists and online bulletin boards, including a suite of email lists called "Independent Living on the Autistic Spectrum" [InLv] created by Martijn Dekker'. Also coming to public notice is the autistic utilization of websites like Second Life – home of the Autism Island, 'Brigadoon'[6] – and YouTube (e.g., Gajilan 2007), and the prevalence of groups like 'Posautive' and 'Autism and Computing' to raise awareness about and foster alternative, affirmative views of autism (Murray and Lesser 1999). Much of the work of advocacy takes place in virtual arenas, and lends weight to the claim that 'the democratization of information flow which is the Internet has promoted the emergence of new ways of self-identification for autistics' (Singer 1999: 64). Proclamations for the Net's significance for autism are in fact often momentous; witness the same author's oft-repeated statement that 'The impact of the Internet on autistics may one day be compared to the spread of sign language among the deaf' (Singer 1999: 67). Several other autistic authors have emphasized the complex connections between Internet usage and what they see as the emergence of a distinctive autistic culture (Dekker 2006, Prince-Hughes 2004).

There are persuasive indications from AS authors that autism is emotionally, spatially – in every sense – *otherwise* (O'Neill 1999), such that recognition of a distinctive culture becomes not only reasonable, but almost inevitable. That is to say, autistic differences in perception and 'processing' tend to involve Other ways of being-in-the-world, separate senses of selves and space that give rise to distinctive cultural experience, and so also, cultural *expression*. While the sense of phenomenally different perception on the spectrum has been explored in depth elsewhere (Davidson 2008c), below we are concerned with expressly communicative aspects of difference and cultural distinction. As one AS author reiterates: '[m]uch like the deaf community, we autistics are building an emergent culture. We individuals, with our cultures of one, are building a culture of many' (Prince-Hughes 2004: 7; see below and Davidson 2008b on D/deaf and AS cultural connections). Such a comment offers a challenge to geographers who have considered people with disabilities as different and occupying 'distorted space' that they must 'endure' and not celebrate (Golledge 1993: 63, quoted in Gleeson 1999: 103).

---

6   Founder John Lester named Brigadoon after a fictional enchanted Scottish village – 'a beautiful place full of magic' and explains that: 'Second Life is a commercial system (you pay a monthly charge to use their services), and although it is often classified as an "online game" (like The Sims Online), it really ISN'T a game. It's an online world, with no content or goals except those created by the people who use it. [...In Brigadoon, AS users can] build a world around themselves within which they could talk to each other and create a whole new kind of online community' (Lester 2005: npn).

For people with mental health problems – a potentially recoverable condition of difference and therefore distinctive from the pervasive permanency of AS – one key impact of the Internet has been an alleged 'democratizition' of medical information, which has empowered individual patients to understand more about particular health conditions in ways potentially disruptive of traditional hierarchies in medical relationships (Parr 2002). Current and new patients can research online their treatments, prognoses and care providers, amongst other things. However, the benefits of Internet technology go beyond examples of individual accumulations of 'expert' knowledge. There are now many examples of new forms of communality shared by such vulnerable people in virtual space, with chat rooms, email lists and forums focused around mutual support and the sharing of information and coping strategies around particular diagnostic categories. Internet users can cross-compare experiences, medications and progress on discussion forums which effectively 'open up' psychiatric bio-medicine to embodied talk and situated knowledges (Wikgren 2001). The search for, finding and reading of mental health-related information is, of course, infused with a politics of knowledge that makes such processes confusing, risky and ambivalent, as well as empowering for Internet users.

In an attempt to negate the dangers of what Haythornwaite and Wellman have called 'networked individualism' (2002: 32) in relation to mental health problems, it is also (along with AS) possible to point to various electronic mental health activisms taking place, which offer collective cultures of resistance to the medical and legal status quo. They take a variety of forms from politicizing the use of certain disabling medication and contentious treatments (see above and www. benzo.org.uk, www.ect.org, www.seroxat.pwp.blueyonder.co.uk)[7] to campaigning about discrimination (www.mindout.clarity.uk.net) and building mental health awareness (www.madnotbad.co.uk, http://groups.yahoo.com/group/uksurvivors), here just giving UK examples. Participation can include direct off-line liaison with MPs involved in particular campaigning issues (see www.benzo.org.uk) in ways that circumnavigate the usual paths of mental health representation through advocacy organizitions. Virtual mental health activisms also enable political participation beyond the nation-state, rendering accessible global networks orientated around human rights and mental health care (http://www.mindfreedom. org), as well as more anarchic celebrations of madness through international 'Mad Pride' networks (http://www.ctono.freeserve.co.uk). This selection of examples represent access routes whereby people with mental health problems might 'act' at a variety of scales from the body to the global – as users of services, or as 'mad' people – in order to effect 'economic, social, environmental and cultural integration of user/survivors into mainstream society and our active participation and integration into community life' (http://www.ctono.freeserve.co.uk). Such online activisms involve different requirements in terms of whether and how individuals collectively identify as patients or users of treatments and services.

---

7 All websites referenced in this chapter were accessed between January and September 2006.

Crossley (2002: 70) has argued that mental health 'movements' have not typically used the activist tactics of the disability movement, partly because disruptive embodied behaviours in public spaces can be read as evidence of irrationalities. Online networking may thus help to articulate less risky activisms in this regard, while still contributing to opportunities to 'rescript illness' scenarios (Moss and Dyck 2003) and disabled subjectivities, so resisting ableism in multiple ways.[8]

Highlighting connections with online activities associated with AS, Clarke and van Ameron (2007) have found that resistance to the 'medicalization' and professional 'ownership' of autism is a common discursive feature of AS sites. Others recognize that 'the use of the Internet by individuals with autism and related conditions is part of a movement of self-advocacy' (Brownlow and O'Dell 2002: 690). At the most basic level, 'the Internet has begun to challenge stereotypes surrounding the competence of people with autism to communicate effectively' (Brownlow and O'Dell 2006: 315). Put somewhat differently, the Internet enables those with AS to have a *voice*, a collective voice that is often confrontational in contesting dominant constructions of autism.

Although, as we demonstrate above, there *are* ways to align the outcomes of an online presence for those with AS and mental health problems, we want to do more than highlight the Internet as a technology enabling of politicized networks, important as this is. Rather, in what follows below we draw out differences in how and why the Internet is used by those with AS and mental health problems in order to show how this is a technology offering access to virtual spaces in which it is not possible to reductively conclude that new orders of normality and deviancy are simply (re)made. Rather, we emphasize that *particular proximities* to disordered differences are evidenced and are suggestive of quite complex geographies of encounter both within these particular groupings and also between these groupings and 'mainstream' others. This evidence also suggests that the impacts of the Internet as a technology are constantly being rearticulated by 'other/ed' (or disabled) groupings in ways that are iteratively designed by their own disordered cultures. Such a claim helps to challenge a disability theory which renders technologies as simply productive of (only) ableist and compensatory logics. Below, we trace these arguments more fully by profiling the distinctive language cultures of AS virtuality as described by AS authors themselves. Here AS is represented as a (celebrated) neurological difference increasingly known and recognized by 'the mainstream' by virtue of the Internet. This is contrasted with research evidence

---

8 For Moss and Dyck (2003), the notion of 'rescripting' emerged in response to the need to conceptualize complex practises engaged in by people with chronic illness to manage their ill identities and experiences. In various, often imaginative and sophisticated ways, such individuals employ the politics and practice of rescripting to contest and re-work their positioning (by the state, family, colleagues and so on), and to construct mind- and body-spaces more conducive to the fluctuating challenges of their condition (see Parr 2008, for elaboration of this concept with specific relation to the experience of those with mental health problems).

from virtual mental health forums which demonstrate an (ironic) interiorism in which proximities to difference are experienced largely by users of the forums themselves, and which involve the individual (re)drawing of social boundaries in order to create distance from abject relations experienced as threatening to a recovering and fragile self. These two different examples provide a way to theorize the potential of the Internet as a disabling/enabling technology of difference.

Communicating Cultures of Autism Online

Perceptual and 'processing' differences associated with autism hinder typical communication, with the result that words and actions of others can be unpredictable, indecipherable and scary for those on the spectrum. In contrast, and as numerous authors attest, AS speech tends to be clear, to the point, and to avoid any reference to extraneous information that might muddy the clarity of communicative intent: 'I always said exactly what I meant, neither more nor less. That other people didn't do that was very confusing' (Gerland 2003: 35, see also Williams 2005). Further emphasizing distinctions between AS and 'neurotypical' (NT) styles of communication, Dekker (2006: npn) writes: 'NT conversations have a very fast-paced rhythm of little exchanges back and forth, whereas autistic people usually say what they have to say, in its entirety, then stop talking and wait for the other to respond.' To function in everyday situations, those with AS describe having to develop skills as sociologists (Dave, in Osborne 2002: 68) and anthropologists to 'figure out the natives' (Grandin, quoted in Sacks 1995: 256), and find ways to fit in. However, the efforts involved with such constant performance in different cultural spaces are exhausting:

> That people with autism have to exist within a different culture on a day-to-day basis in order to survive – one that often blindly insists on conformity rather than respecting our cultural diversity – makes functioning in the world around us exceedingly difficult, often depressing and continually anxiety-laden (Grandin and Barron 2005: xvi).

It is small wonder that people with AS seek out alternative spaces where proximity with NTs and pressure to pass is absent, where communication can take place according to their own precisely straightforward rules of engagement. Autistic spoken communicative styles have been described as 'comparable to written communication' (Dekker 2006: npn), and so the significance of text is unsurprising. The use of letter, then fax, now computer, are common in AS accounts, and for such as Williams, type-writing provides a valuable means of clarifying thoughts; 'When I need to explain something at a level of complexity for which spoken words evade me, I still run off to the computer and let my fingers talk' (Williams 2005: 252). Computational and mechanical predictability is suited to AS challenges in many ways. As Shore (2001: 60) explains:

computers are often particularly well suited for those on the autistic spectrum as they provide interactive consistency. A computer has the same response for a given input, so there is no body language or tone of voice messages that need to be decoded.

Communication with others via computer for those with AS is thus in many ways ideal, satisfyingly straightforward and clear when compared to sensorially overwhelming face-to-face encounters. As Nelson explains (2004: npn), comfortable proximities enable the emergence of 'far-flung' autistic *communities*, and so virtually disassembles at least some disabling geographies:

> People on the autism spectrum have a unique social network, this is primarily using communication with text on the Internet. It is an invaluable community for many of us. There should be increased availability and recognition for this autism community online so that isolated members of the autism community can join and participate.

Clearly, the Internet is crucial for challenging isolation and exclusion amongst members of a group who need spaces of solitude, but desire social contact. The medium can itself be a uniquely important 'meeting place' (Nelson 2004: npn) in surprisingly substantial ways. The following quotation is drawn by Singer (1999: 65) from an Independent Living (InLv) thread about the ideal 'country' for autistics: 'We've already got our own country. It's a cybercountry called InLv, and it's perfect. We can interact without getting on each other's nerves – gently, carefully'. Such social space online might thus be in some senses Utopian – a literal 'no-place' where subjugated and 'disabled' identities can be cast in a different light. 'Here', it is not only that pressures to be 'normal' disappear, but that differences are experienced communally, 'abnormalities' shared, in and on AS authors' own terms. In place of perpetual isolation, AS subjectivities online are reiterated or rescripted *in relation* and experienced in more positive, even celebratory terms. By being virtually proximate with similarly different others, those with AS can perform and practice emotional relatedness, allowing new understandings and cultures to emerge that are not concerned with becoming normalized, and that actively challenge ideas of incapacity and deviancy.

Clarke and van Ameron report that AS bloggers 'spoke of celebrating their differences and of anger at neurotypicals for stigmatizing them'; that they are 'proud of their differences and committed to the value of the uniqueness and the nuances of their experiences of AS' (2007: 771-772).[9] AS authors increasingly assert minority status based on collective communicative culture and identity: 'With our own communication medium, autistics are beginning to see ourselves not as blighted individuals, but as a different ethnicity' (Singer 1999: 67).

---

9   This study, however, has found that very few of those writing about AS do so in ways that are straightforwardly or consistently celebratory (Davidson 2008a, 2008b).

Important and enabling as this positive sense of difference clearly is, one might anticipate concerns about an associated potential for cultural *separatism*, and perhaps even 'ghettoization'. However, despite many challenges with cross-cultural communication, no AS author seems to advocate anything other than multiculturalism as the most appropriate path for autistic 'inclusion'.

Referring to the expansion of AS horizons beyond the borders of AS culture, Blume quotes one respondent's claim that: 'The level of communication possible via the Internet is changing our lives, ending our isolation, and giving us the strength to insist on the validity of our own experiences and observations' (in Blume 2007b: npn). Crucially, challenges with cross-cultural communication in real time and space are addressed; 'Of those autistics on the Internet who discuss its use, we all agree that its [sic] an amazing tool [...] because of the way that it allows for a delay in a response that is almost never allowed in real life' (Camille Clark, quoted in Biever 2007: npn). In the real world, explains Darius (2002: 25), 'There is no such thing as adequate delayed social reactions. One either is quick enough to keep up, or one is weird and socially disabled'. In stark contrast to such disabling real-time and space exchange:

> On the Internet, freed from the constraints of NT timing, NT ways of interpreting body language, free from the information overwhelm of eye contact, the energy demands of managing body language, [those with AS] sound, simply, 'normal', and often, eloquent (Singer 1999: 65).

Here, however, we encounter potential cause for concern of the kind anticipated by Gleeson (1999) and Moser (2006), where the Internet might begin to be seen as an inappropriate technological fix intended to 'overcome' the limitations of a group labelled disabled, by enabling a performance of normality. Whereas some of those with autism compare the Internet to sign-language for those who are D/deaf, the quote above might be taken to suggest that its function is more akin to cochlear implants, a 'fix' that some argue attempts to erase rather than admit sensory (and cultural) difference (see Valentine and Skelton 2003, Davidson 2008b). While there is considerable evidence to suggest that the Internet facilitates a degree of social inclusion, allowing productive cross-cultural exchanges to take place more easily, and simply *more*, than ever before, we have to question on whose terms and for what purpose such exchanges take place. Concerns might be raised, for example about degrees of *real* socio-spatial inclusion which result.

New research is certainly needed to explore the potential restrictions as well as capacities that go along with Internet use. Nonetheless, important evidence is emerging that suggests AS voices 'gathering force' online *will be heard off-line too*, that virtual communication has the potential to spill over into the 'real world', with further potential for political consequence. As one AS individual explains: 'When the computer became able to connect me with others via the Internet, my 'real' world expanded also' (in Blume 1997a: npn). Shore (2001: 142) emphasizes opportunities to travel and socialize while 'spreading the word' about autism:

'Cyberspace can be a good place for those on the autistic spectrum to meet others. For example, I have been invited to present at several conferences as a result of my cyberspace connections'. Dekker (2006: npn) also explains how the positive space of a virtual community takes on real world presence through 'Autreat', a three-day conference camp in New York State 'that replicates the autistic space in 3D life'. AS cultures emerging online thus have the potential for real world outcomes, and indeed recent discussion at Autreat 2008 focused on the practicalities, politics and potential benefit of establishing a physically situated autistic community (Orsini, pers comm.[10]). Clearly, spaces of, and for, autistic identities will continue to change in both online and off-line worlds, and will do so in ways that impact on, and perhaps further *connect* with, so-called mainstream society.

Mental Health and New Proximities of Difference

> Sometimes I think if it wasn't for the Internet I wouldn't have even developed the social skills that I have. My phobias (social phobia mainly) has stopped me from developing 'in that way' in real life – but the Internet has stimulated my development instead (female, NPS[11]).

There is little doubt that there are new and extraordinary benefits that virtual environments bring to people with severe mental health problems in a disabling society in much similar ways to the (celebratory) visioning of AS virtuality reported above. However, in-depth research (see Parr 2008, Chapter 6) on the uses and impacts of mental health support forums reveals *particularities* about the experience of online communality for those with mental health problems and has ramifications for our thinking on the orders of normality and difference that are constructed via technologies for, and amongst, disabled groups.

Research (Parr 2008) shows that one value of the experience of online co-writing is a sense of expressiveness and positive release of feelings hidden in a 'sane', disabling, off-line society:

> When I'm online I know I can express myself, reveal my innermost thoughts and feelings without fear. I can say things online that I could never say in everyday conversation (female, MHUK).

In forums for online mental health support and discussion, disruptive emotions, thoughts and behaviours can make up significant amounts of the content of virtual talk and socializition. Postings often detail current emotional states and

---

10   Michael Orsini (School of Political Studies, University of Ottawa) is research collaborator with Davidson on a SSHRC-funded project exploring autistic activism and identities in online and off-line space.

11   See footnote 2 for details of the study from which these materials are drawn.

also ask for responses, advice or solutions to difficult situations. The dense writing of disruptive thoughts and emotions exposes other online users to often quite difficult intimacies, especially if posted from regular members of forums, because the contents of posting must be noted and addressed in ways that often challenge notions of technological 'drive-by relationships' (Puttnam 2000: 177). The Internet for mental health support is hence reported as being an intense space and one used for a variety of reasons – to offload, express, ask for help and practise emotional and social relatedness (often disrupted during periods of illness). In this latter regard there are comparisons to AS virtual cultures, where computer communication offers a less problematic means of co-encounter than the social occupation of off-line spaces. Yet, whereas some people with AS speak of the uncomplicated territories of online talk, there is evidence that for people with mental health problems, muddy, nuanced and un-clear emotional talk can at times threaten fragile psycho-social boundaries and thus seem to offer less straightforward communicative spaces than first might be suggested.

What does this mean for proximities to, and of, mental health differences? Amongst users of support forums the intensive exposure to the detailed writing of the disordered self that the Internet brings has particular effects. The particular temporality of e-relationships, for example, demands a certain 'instantness' suggestive of temporal, if not spatial, proximity. The expectations of users for responses at all times of the day and night is one of the key named benefits of Internet forum use, although when responses are not immediately forthcoming this can lead to negative emotions and even erosions of trust. Gwinnell (2003: 329) also argues, for example, that 'constant checking for new email becomes part of any Internet relationship. The cycle of anticipation, fantasy, anxiety, relief from anxiety and new anticipation seems to produce obsessive-compulsive behaviour patterns'. Although such a pathological reading of temporal e-proximities is problematic, the suggested emotional reliance on written responses is nonetheless accurate, as Gilly (telephone interview, MHF) says: 'I was so disappointed that I put on a posting you know and I got, I think got one reply two days later. And that has actually put me off putting on anything emotional, not getting any response.' That some users may also find it difficult to put others 'on hold' as a result of these stated feelings suggests that supportive e-relationships carry new responsibilities and pressures for participants, and this might prove difficult for those dealing with mental health problems.

Intimate feelings that are often connected to mental health issues and that are shared via graphic online writings undoubtedly help to construct risky emotional spaces. Needless to say, many forum users find dealing with this 'profound horror' (Radin 2006: 594) challenging: '…it can also be dangerous. Some subjects can trigger a bad reaction in me, so I have to be careful to avoid those areas of forums sometimes' (female, MHF). The graphic representation of subjective irrationalities in online communities fosters new kinds of uncomfortable proximities to suffering that may render it difficult to respond to postings even from members with whom trust and good relations are already established. The content of the responsive-posting may also prove challenging in terms of both what is said and whether it adequately demonstrates attention to detail and content, something that often worries forum members:

> …you feel you've, you've then got to, to say something that's supposed to help the person, but actually you've got to be so careful about what you say. I know if I'm not well, and somebody just says just two words that are wrong it just sets me off… (Gilly, telephone interview, MHF).

It is easy to imagine how fragile senses of communality might breakdown for a vulnerable person waiting for responses to a sensitive posting. For some members who may not feel well enough to 'do' constant, sensitive and responsive relationship work, these challenges threaten the sustainability of virtual mental health forums. Occupying online forums as places to socialize, support and be supported is certainly an ambivalent endeavour, as some 'do feel part of an on-line community, but whether I feel comfortable there is still to be decided' (female, NPS). Accumulated self-writing on forums can thus be uncomfortable, distressing and even dangerous, although it *also* provides users with complicated opportunities through which to *practice* self and social management. Some users do this through diverse acts of distancing: 'I always seem to end up giving more support than I receive – which is why I "cut and run" quite often' (male, NPS). In this regard management tactics include leaving forums, breaking contact, filtering or deleting messages, simply sending virtual 'hugs' without in-depth replies, recommending off-line medical contact and appealing to the forum moderators and/or website rules. That user-led online forums provide opportunities to practice different forms of boundary maintenance with peers is potentially useful, and also signals that people with mental health problems can exercise competencies in managing relationships (a social 'norm'). However, although such distancing strategies do not *always* suggest that intimate virtual proximities are productive of *new* forms of exclusions within forums, there is still a possibility that some writers may be *too* irrational, *too* different for particular forums (and various examples of literal exclusions from support forums are constantly evidenced across online communities of various hues). The particularities of textual relations between people suffering traumatic psychological experiences, then, mean that some boundary maintenance is thought necessary to ensure the sustainability of long-term support, precisely because of the (relative) intimacy of social relations here:

> [the forum] is a place for support, friendship and care, a place to share experiences, to gain understanding. There of course are hard times for members, who struggle day to day with living, in distress and need a place like MHUK to release emotions, to vent…Boundaries are important, not only in the caring field as a professional, but in friendships too and overstepping those boundaries can lead to disaster. I don't want to sound blunt and uncaring, but when I first started MHUK, I was in the very position of feeling and thinking I could hold everyone

together, it soon became apparent that I could not cope with doing that and so learned to 'step back' and logically try and get the best help I could for people without thinking I had somehow destroyed them in the process (Moderator statement, MHUK).

The proximities to, and of, difference that are facilitated and co-produced via the Internet as a technology are not overtly serving as an engine of political or social movements (as in the example of AS) for the majority of people with mental health problems (although this does occur for some, see references above). These proximities are instead rather contentious ones that seem to incur risk and the risk of abject relations between psychologically vulnerable co-writers. That these abject relations may sometimes result in routine boundary maintenance (Sibley 1995) potentially has implications for our thinking about the virtual constructions of normality and deviance as people with mental health problems use Internet technologies for their complex recovery work. This recovery work may involve practicing social interaction in specific mental health support forums and then 'moving on' to interactions in 'mainstream' online forums or off-line spaces as we saw with AS (Parr 2008). One risk of this mobility is that the 'depth' emotional talk of mental health forums is constructed as 'abnormal' and unlike other forms of virtual talk in other online places. Here distinctive communicative norms can be used by those recovering from illness experiences as examples of psychological deviance, and thus notions of communicative normality may emerge even amongst (former) users of support forums.

> I was looking through my emails, my sent emails on the Internet...my emails to people [when] I was actually quite manic (Mandy, telephone interview, NPS).

> It is easier, for obvious reasons, to communicate with non-phobic/mentally ill people. I have had a few friends from the phobic society, but they all were more ill than me and eventually I had to break contact or be brought down by them (male, NPS).

Such evidence complicates claims about disabling barriers being (potentially) removed by virtual communality, and suggests that nuanced constructions of normality and difference are in operation even *amongst* those disabled by mainstream attitudes to mental health problems. There may be lessons here for the emerging virtual movement of AS people, and claims for distinctive virtual cultures of communication, or *more likely*, it may be that there are quite different issues at stake for these groups, a point we pursue more directly in the conclusion of this chapter.

Conclusions: Expanding Horizons of Difference?

This chapter has shown that the Internet is experienced as an appropriate and unusually accommodating medium for those with AS, given characteristic preferences for communication at a socio-spatial (and minimal temporal) distance. It has also, however, revealed, through examination of online experiences of those with mental health problems, that the Internet can feel dangerous, and be encountered as a disturbingly intimate psycho-social space. Such dangers have not (yet) been reported by those with AS. In considering such points of comparison and divergence between current (and potential future) experiences of the two 'groups' discussed, we return to thinking about the Internet as a technology in which difference is configured in complex ways with various outcomes, and constituted responsively to particular cultures of 'disorder'.

The Internet is a technological space that facilitates the recognition of difference; individuals and groups can 'work' on rescripting difference textually in order to forge its co-recognition between those implicated and other majorities. *How* this difference is constituted through this technology clearly varies in the examples we have used above. On the one hand, for those with AS, this is a technology that is facilitating voice/s and the forging of distinctive AS cultures. Interestingly, the very socio-spatial distance that attracts AS individuals to Internet communication is that which is rendering them temporally proximate in networked movements, ones full of the promise of positive cultural and (perhaps) disabled difference. Distance and proximity work in other ways for people who use mental health forums, and we have shown above how the Internet can result in up-close, uncomfortable and abject spaces of risk. Unlike celebratory visionings of AS virtual cultures, we deliberately play upon the rather risky interiority of support forums which sometimes result in new formulations of same/other for some of those recovering from mental health problems, and who then remove themselves from the daily round of ill textual intimacy. Here we have the contrast in a nutshell: the Internet can exist as a technology which facilitates space for the difference of AS through the forging of new online cultures of a neurological minority but *also* as a space of recovery from difference in the case of those with mental health problems. That is to say, for many of those with AS who were socially isolated off-line, the Internet facilitates *relational* rescriptings of disorder into difference, difference that becomes accepted, even celebrated, such that notions of 'recovery' are actively rejected. In both cases, it is possible to use the Internet to practice emotional relatedness, gain support, and find information. Differential possibilities for rescriptings and recoveries serve to highlight how the Internet can offer disabled, disordered and otherwise 'different' people genuinely diverse outcomes, which suggests it is a technology resistive of labels of ableism and compensatory logics (cf. Moser 2006).

To elaborate this point further, we can argue that the Internet helps construct spaces of difference and spaces of recovery for different disabled/excluded groups. Neither scenario means that the technology is necessarily compensating for disabled difference. In the case of AS, full recovery is not an option, nor required or desired. And so, how do the iterative textual cultures of Internet communication have meaning here beyond merely connecting people with the same neurological difference for social or political purposes? How can one 'move on' when disorder/difference is integral, pervasive and can never be left behind? To some extent, these questions have still to be answered (see Davidson and Orsini 2008) although in the examples above there is a suggestion of experiential appreciation of conversations, paced virtual talk, friendships and shared dwelling spaces that have potentially significant cultural implications. In this sense a recovery from disabling isolation is possible. For people with mental health problems of various hues the Internet can also provide a sociable place of refuge as well as a space for practicing 'normal' emotional and social functioning. For many long-term users of virtual mental health support forums, 'normality' is a contested term, but nonetheless people use the Internet to help them 'be' and 'pass' in mainstream, off-line spaces and relationships. Considering the increasing use of Internet technology for social networking in mainstream cultures (Haythornwaite 2005) such a function does not suggest that this is wholly particular to this group. This is a technology which can arguably recast what we understand of disabled difference in respect to mental health problems and AS cultures, especially as the sociality of users of specific condition-related sites and forums begins to 'bleed' into other mainstream social networking hubs.

So what of the future of disabled difference and Internet technology? This is perhaps too grand a question for the short chapter above, but we can begin to offer some thoughts about the implications that may lie in store for those specific groups we have researched above:

> autistics are constituting themselves as a new immigrant group on line…in trying to come to terms with an NT-dominated world, autistics are neither willing nor able to give up their own customs. Instead, they are proposing a new social contract, one emphasizing neurological pluralism (Blume 1997a: npn).

While it seems that, to date, the place of autistic activism remains largely online, many are actively contributing to the creation of a collective voice and 'movement', one that contests predominant constructions of AS difference as disorder or disability and one that argues for the social inclusion of 'new' neurological minorities. Our research has also shown that those with various and often recoverable mental health problems can benefit from practicing social relations and re-establishing boundaries online – despite the risks of highly emotive and often muddy communications – in a way that has potential to make negotiating mainstream space more manageable, even with significant psychological and emotional difficulties which may still be present. Serious questions must remain of course about such a scenario, and caution must still be exercised over any reductive technological reasoning, so as to avoid what Gleeson (1999) identifies as the 'rehabilitative' forces of technological 'antidotes' to disabled difference:

There is nothing wrong with disabled people that the proper environment can't fix...Technology can solve anything...the problem is getting people to use devices (Scherer 1993: 84, quoted in Gleeson 1999: 99).

We are not suggesting above that the Internet can simply mediate (or remediate) difference, but rather argue that it can be enrolled in various 'rescripting' scenarios (Moss and Dyck 2002) for both AS and mental health problems, albeit that the grounds of that rescripting may be differently configured in each case. As new communications technology opens new spaces for experiencing and expressing difference, we have to be prepared to embrace its diverse potential and not just critically consign it to a limiting mechanical fix.

Acknowledgements

Empirical materials from the two studies presented in this chapter have previously been discussed in Parr (2008) *Mental Health and Social Space: Towards Inclusionary Geographies*, and Davidson (2008) Autistic cultures online: virtual communication and cultural expression on the spectrum. *Social and Cultural Geography*, 9(7): 791-806. We would like to thank Blackwell Publishing and Taylor and Francis, respectively, for permission to reproduce some of this material in its current form.

References

Aarons, M. and Gittens, T. 1999. *The Handbook of Autism: A Guide for Parents and Professionals*. London and New York: Routledge.

American Psychiatric Association. 2004. *Diagnostic and Statistical Manual of Mental Disorders*. Washington, DC.

Avrahami, E. 2007. *The Invading Body: Reading Illness Autobiographies.* Charlottesville: University of Virginia Press.

Bagatell, N. 2007. Orchestrating voices: autism, identity and the power of discourse. *Disability and Society*, 22 (4), 413-426.

Baker, D.L. 2006. Neurodiversity, neurological disability and the public sector: notes on the Autism Spectrum. *Disability and Society*, *21*(1), 15-29.

Biever, C. 2007. Web removes social barriers for those with autism. *New Scientist* 2610. [Online, 27 June] Available at: http://technology.newscientist.com/article/Mg194261106.100 [accessed: 27 April 2009].

Biklen, D. (ed.) 2005. *Autism and the Myth of the Person Alone.* New York: New York University Press.

Blume, H. 1997a. Autistics, freed from face-to-face encounters, are communicating in cyberspace. *The New York Times.* [Online, 30 June] Available at: http://query.nytimes.com/gst/fullpage.html?res=9803E7DC1F31F933A05755C0A961958260&sec=&spon=&pagewanted=all [accessed: 21 January 2008].

Blume, H. 1997b. 'Autism and The Internet' or 'It's The Wiring, Stupid'. Media In Transition, Massachusetts Institute of Technology. [Online] Available at: http://web.mit.edu/m-i-t/articles/index_blume.html [accessed: 21 January 2008].

Blume, H. 1998. Neurodiversity: on the neurological underpinnings of geekdom. *TheAtlantic.com.* [Online] Available at: http://www.theatlantic.com/doc/199809u/neurodiversity [accessed: 21 January 2008].

Brownlow, C. and O'Dell, L. 2002. Ethical issues in qualitative research in online communities. *Disability and Society*, 17(6), 685-694.

Brownlow, C. and O'Dell, L. 2006. Constructing an autistic identity: AS voices online. *Mental Retardation*, 44(5), 315-321.

Clarke, J. and van Amerom, G. 2007. 'Surplus suffering': differences between organizational understandings of Asperger's syndrome and those people who claim the 'disorder'. *Disability and Society*, 22(7), 761-776.

Cresswell, T. 1996. *In Place, out of Place: Geography, Ideology and Transgression.* Minneapolis and London: University of Minnesota Press.

Crooks, V.A. 2006. 'I go on the Internet; I always, you know, check to see what's new': Chronically ill women's use of online health information to shape and inform doctor-patient interactions in the space of care provision. *ACME: An International E-Journal for Critical Geographies*, 5(1), 50-69.

Crossley, N. 2002. Repertoires of contention and tactical diversity in the UK psychiatric survivors; movement: the question of appropriation. *Social Movement Studies*, 1, 47-70.

Darius, 2002. In *Aquamarine Blue 5: Personal Stories of College Students with Autism*, edited by D. Prince-Hughes. Athens, OH: Swallow Press, 9-42.

Davidson, J. 2003. *Phobic Geographies: The Phenomenology and Spatiality of Identity*. Burlington VT and Aldershot: Ashgate Press.

Davidson, J. 2007. 'In a world of her own...': re-presentations of alienation in the lives and writings of women with autism. *Gender, Place and Culture*, 14(6), 659-677.

Davidson, J. 2008a. 'More labels than a jam jar...' the gendered dynamics of diagnosis for girls and women with autism, in *Contesting Illness*, edited by P. Moss and K. Teghtsoonian. Toronto: University of Toronto Press, 239-258.

Davidson, J. 2008b. Autistic culture online: virtual communication and cultural expression on the spectrum. *Social and Cultural Geography*, 9(7), 791-806.

Davidson, J. 2008c. Feeling different: sensory geographies of autism. Under review. Copies available from author.

Davidson, J. and Orsini, M. 2008. *Autism Online: The Social and Cultural Implications of the Internet for Individuals on the Spectrum*. Proposal approved by the Queen's University General Research Ethics Board (19 August).

Davidson, J. and Parr, H. 2007. Anxious subjectivities and spaces of care: therapeutic geographies of the UK National Phobics Society, in *Therapeutic Landscapes*, edited by A. Williams. Burlington, VT and Aldershot, UK: Ashgate, 95-110.

Dekker, M. 2006. On our own terms: emerging autistic culture. *Autistic Culture.* [Online] Available at: http://autisticculture.com/index.php?page=articles [accessed: 21 January 2008].

Eng, R.T., Maxfield, A., Patrick, K., Deering, M.J., Ratzan, S.C. and Gustafson, D.H. 1998. Access to health information and support: a public highway or a private road? *The Journal of the American Medical Association*, 280(15), 1371-1375.

Fairclough, N. 1995. *Critical Discourse Analysis: The Critical Study of Language.* London: Longman.

Fox, N. 2001. Use of the Internet by medical voluntary groups in the UK. *Social Science and Medicine*, 52, 155-156.

Gajilan, A.C. 2007. Living with autism in a world made for others. [Online] Available at: http://www.cnn.com/2007/HEALTH/02/21/autism.amanda/index.html [accessed: 7 January 2008].

Gerland, G. 2003. *A Real Person: Life on the Outside.* London: Souvenir Press.

Gevers, I. 2000. Subversive tactics of neurologically diverse cultures. *The Journal of Cognitive Liberties*, 2(1): 43-60.

Giggin G and Newell C. 2006. Editorial comment: disability, identity and interdependence: ICTs and new social forms. *Information, Communication and Society*, 9(3), 309-311.

Gleeson, B. 1999. Can technology overcome the disabling city? in *Mind and Body Spaces: Geographies of Illness, Impairment and Disability*, edited by R. Butler and H. Parr. London: Routledge: 98-118.

Grandin, T. and Barron, S. 2005. *Unwritten Rules of Social Relationships: Understanding and Managing Social Challenges for Those With Asperger's/ Autism.* Arlington, TX: Future Horizons.

Grinker, R.R. 2008. *Unstrange Minds: Remapping the World of Autism.* New York: Basic Books.

Hahn, H. 1989. Disability And the reproduction of bodily images: the dynamics of human appearances, in *The Power Of Geography: How Territory Shapes Social Life*, edited by J. Wolch and M. Dear. Unwin Hyman: London, 370-388.

Haythornwaite, C. 2005. Social networks and Internet connectivity effects. *Information Communication and Society*, 8(2), 125-147.

Haythornwaite, C. and Wellman, B. 2002. *The Internet in Everyday Life.* Oxford: Blackwell.

Iverson, P. 2008. *Strange Son.* New York: Basic Books.

Jones, R.S.P., Zahl, A. and Huws, J.C. 2001. First-hand accounts of emotional experiences in Autism: a qualitative analysis. *Disability and Society*, 16(3), 393-401.

Lawson, W. 2005. *Life Behind Glass: A Personal Account of Autism Spectrum Disorder.* London and Philadelphia: Jessica Kingsley Publishers.

Lester, J. 2005. About Brigadoon. [Online] Available at: http://braintalk.blogs.com/Brigadoon/2005/01/about Brigadoon.html [accessed: 7 January 2008]

Madge, C. and O'Connor, H. 2006. Parenting gone wired: empowerment of new mothers on the Internet? *Social and Cultural Geography*, 7, 199-220.

Moser, I. 2006. Disability and the promises of technology: technology, subjectivity and embodiment within the order of the normal. *Information, Communication and Society*, 9(3), 373-395.

Moss, P. and Dyck, I. 2003. *Women, Body, Illness: Space and Identity in the Everyday Lives of Women with Chronic Illness*. Lanham, Maryland: Rowman and Littlefield.

Murray, D. and Lesser, M. 1999. Autism and Computing. [Online] Available at: http://autismandcomputing.org.uk/computing.en.html [accessed: 7 January 2008].

Nelson, A. 2004. Declaration from the Autism Community that they are a minority group. *PRWeb: Press Release Newswire*. 18 November [Online]. Available at: http://www.prweb.com/releases/2004/11/prweb179444.htm [accessed: 21 January 2008].

O'Neill, J.L. 1999. *Through the Eyes of Aliens: A Book About Autistic People*. London: Jessica Kingsley Publishers.

Osborne, L. 2002. *American Normal: The Hidden World of Asperger's Syndrome*. New York: Copernicus Books.

Parr, H. 2008. *Mental Health and Social Space*. Oxford: Blackwell.

Parr, H. and Davidson, J. 2008. 'Virtual trust': online emotional intimacies in mental health support, in *Researching Trust and Health*, edited by J. Brownlie, A. Greene and A. Howson. New York and London: Routledge, 33-53.

Prince-Hughes, D. 2004. *Songs of the Gorilla Nation: My Journey Through Autism*. New York: Harmony Books.

Purkis, J. 2006. *Finding a Different Kind of Normal*. London and Philadelphia: Jessica Kingsley Publishers.

Putnam, R.D. 2000. *Bowling Alone: The Collapse and Revival of American Community*. Toronto: Simon and Schuster.

Radin, P. 2006. 'To me, it's my life': medical communication, trust, and activism in cyberspace. *Social Science and Medicine*, 62, 591-601.

Sacks, O. 1995. *An Anthropologist on Mars*. Picador: London.

Segrott, J. and Doel, M. 2004. Disturbing geography: obsessive-compulsive disorder as a spatial practice. *Social and Cultural Geography*, 5(4), 597-614.

Shore, S. 2001. *Beyond the Wall: Personal Experiences with Autism and Asperger Syndrome, Second Edition*. Shawnee Mission, Kansas: Autism Asperger Publishing Company.

Shore, S. 2006. The importance of parents in the success of people with autism, in *Voices from the Spectrum*, edited by C.N. Ariel and R.A. Naseef. London and Philadelphia: Jessica Kingsley Publishers, 199-203.

Sibley, D. 1995. *Geographies of Exclusion*. London and New York: Routledge.

Singer, J. 1999. 'Why can't you be normal for once in your life?' From a 'problem with no name' to the emergence of a new category of difference, in *Disability Discourse*, edited by M. Corker and S. French. Buckingham: OU Press, 59-67.

Smith, S. 1996. Taking it to a limit one more time: autobiography and autism, in *Getting a Life: Everyday Uses of Autobiography*, edited by S. Smith and J. Watson. Minneapolis: University of Minnesota Press, 226-248.

Smith, S. and Watson, J. 2001. *Reading Autobiography: A Guide for Interpreting Life Narratives.* Minneapolis: University of Minnesota Press.

Szatmari, P. 2004. *A Mind Apart: Understanding Children with Autism and Asperger Syndrome.* New York and London: The Guilford Press.

Valentine, G. and Skelton, T. 2003. Living on the edge: the marginalization and resistance of D/deaf youth. *Environment and Planning A*, 35, 301-321.

Waltz, M. 2005. Reading case studies of people with autistic spectrum disorders: a cultural studies approach to disability representation. *Disability and Society*, 20(4), 421-435.

Wikgren, M. 2001. Health discussions on the Internet: a study of knowledge communication through citations. *Library Information Science Research*, 23, 305-317.

Williams, D. 2005. *Autism: An Inside-Out Approach: An Innovative Look At the Mechanics Of Autism and Its Developmental Cousins.* London and Philadelphia: Jessica Kingsley Publishers.

Wooten, R., Yellowlees, P. and McLaren, P. 2003. *Telepsychiatry and E-Mental Health.* London: Royal Society of Medicine Press Ltd.

Chapter 5

'It's my umbilical cord to the world…the Internet': D/deaf and Hard of Hearing People's Information and Communication Practices

Tracey Skelton and Gill Valentine

Introduction

Information and Communication Technologies (ICT) have been trumpeted as having the potential to facilitate marginalized social groups to overcome disadvantage (e.g., Foley 2004, Dobransky and Hargittai 2006). These technologies are regarded as particularly important because in an Information Age access to information is a fundamental tenet of citizenship – necessary for engagement with public services, civic participation, social inclusion, and lifelong learning (e.g., European Council, 1994). In particular, a number of studies (e.g., Dobransky and Hargittai 2006, Moser 2006) have focused on the potential role that these technologies may play in undoing disabilities, given that disabilities are now widely acknowledged to be a product of social barriers and not an inevitable outcome of bodily differences (e.g., Oliver 1984, Gleeson 1999). In Moser's (2006: 373) words disabled 'is not something one is but something one becomes'. However, Dobransky and Hargittai (2006) warn against assuming that disabled people's use of the Internet is always related to impairment, and of the dangers of collapsing different types of disability into one category. As such, they call for more in-depth studies into the ICT experiences of different disabled groups. Here, D/deaf, and to a lesser extent hard of hearing, people represent an important, though complex, case in point.

There are over nine million D/deaf and hard of hearing people in the UK. They constitute a very heterogeneous group, made up of people who self-identify in different ways and have different communication strategies. The popular conception of D/deaf people is that they are disabled. Indeed, most D/deaf people are born into hearing families so their first recognition of deafness is usually through a medical model of disability in which the emphasis is on teaching them to communicate orally through lip-reading and speaking and to use technologies (e.g., hearing aids and cochlear implants) to facilitate their integration into hearing society. However, Deaf people whose first, or preferred language, is sign language, regard themselves as a linguistic and cultural minority, arguing that if hearing people learnt to sign then Deaf people would not be 'dis-abled'. Here Deaf is written with a capital D to indicate this construction of a Deaf cultural and linguistic identity. Of course, the boundaries between what are dubbed 'big D' and 'little d' identities can be fluid over time and space. For example, learning sign language often results in a shift in an individual's self-identity from deaf to Deaf (see Skelton and Valentine 2003a for a fuller discussion of D/deaf identities). Here, we use the convention of writing D/deaf in a dual form to reflect this fluidity and complexity, and to render our discussion inclusive of different positionalities. The terms Deaf or deaf are used to refer to the specific differentiated meanings outlined above. Hard of hearing people have historical and cultural backgrounds that are distinct from D/deaf people in that they have usually been hearing but for a range of reasons become deaf. Their communication needs may differ from D/deaf (e.g., needing amplification of sound devices such as hearing loops), although there is increasing recognition of the strategic advantages for D/deaf and hard of hearing people to address common experiences of exclusion from the hearing world. For this reason this chapter addresses the information and communication implications of living in a hearing society that are shared by all those with a hearing impairment regardless of their self-identifications, while also drawing out some of the specificities in the experiences of those who identify as either Deaf or deaf. For example when referring to particular issues relating to sign language use we use the term Deaf.

Communication technologies have historically isolated D/deaf and hard of hearing people from information in mainstream society, for example, the telephone, radio, and television are all inaccessible to D/deaf without relay services or subtitles (although SMS messaging through mobile technologies is an accessible medium). We begin by examining the consequences for D/deaf people's personal information landscapes of their exclusion from traditional orally-based off-line sources of information (e.g., in relation to entertainment, consumption, learning, civic and political participation and so on) because of their hearing impairment and language needs. The chapter then examines how D/deaf people are using the Internet, how on-line practices are affecting their information capacities and social relations, and considers some of the constraints on D/deaf people's abilities to make full and effective use of this information and communication resource. In doing so, the chapter outlines the complex effects of Internet use for different groups of users. Through this analysis we reflect on the concept of 'integration' (cf. van de Ven et al. 2005) and the extent to which the Internet is merely enabling D/deaf people to participate in hearing society or whether D/deaf people's use of the Internet is also challenging and changing hearing 'norms' both on-line and off-line. We conclude with a consideration of possible regulation that might be necessary to enforce the accommodation of D/deaf people, and other disabled people and minority language users', needs more fully.

The material presented here is based on an Arts and Humanities Research Council funded project involving a scoping survey and qualitative interviews. The survey was designed to explore D/deaf and hard of hearing people's access to, and use of, the Internet, to collect data about the informants' self identifications and preferred modes of communication, and was made available on-line. Information about the survey, appeals for people to complete it, and instructions about how to access it were posted on 20 D/deaf related websites. Hard copies of the survey for non-Internet users were distributed to 174 Deaf Clubs/organizations in the UK. In addition a researcher visited specific Deaf clubs/societies across the UK where she was able to introduce the survey face-to-face and answer queries about its completion. Given acknowledged low levels of literacy within D/deaf communities as a result of educational disadvantage (Watson et al. 1999) and research fatigue amongst this hard-to-reach group, the response rate of 419 represents a significant evidence base within the field of Deaf studies. Following the analysis of this survey 42 respondents were recruited to take part in interviews. These interviewees were purposively selected in relation to the patterns identified from the survey. At the time of the research 26 of the respondents defined themselves as Internet users and 16 as non-users. However, we recognize the fluidity of these categories: some non users may become users, some users may abandon this technology, and individuals' actual levels and patterns of usage may vary considerably over time (Seymour 2005).

A hearing researcher fluent in British Sign Language (BSL) conducted the majority of the interviews, and a small number were conducted by Skelton with the support of qualified BSL interpreters. Deaf people have often had to use a variety of forms of communication growing up in a hearing society. BSL is a complex visual language with its own syntax and grammar that differs from oral-aural English (Kyle and Woll 1983). Other sign languages are also used by people with hearing loss/impairments in various language contact situations, including Sign Supported English (SSE) (which was designed to represent English manually). Some deaf people lip-read and speak. The interviews were therefore conducted in the communication mode of each informant's choice.

D/deaf People's Information Needs

All advanced industrialized western states (and many less industrialized nations) legitimize particular ways of communicating in public in which verbal language is privileged over visual and gestural languages, even though 'the use of visual forms of communication is as much part of the natural heritage of human beings as the spoken word...[and] throughout history there are many examples of groups who have developed this ability to communicate visually' (Miles 1988: 8).

The privileging of verbal language matters because it affects the production, consumption and circulation of information. Hearing people are often oblivious to the freedoms and entitlements that come from the ability to pick up information through everyday acts of listening to television, radio, public announcements and so on; as well as through talk in everyday spaces. Yet, this information is difficult for D/deaf people to access because they do not have the proficiency to communicate effectively in oral English (many do not speak, and lip-reading is difficult and often unreliable). Hearing people rarely have signing skills and often lack the patience to try and communicate with non-speaking D/deaf people through gesture or improvised forms of communication (Valentine and Skelton 2007)[1] as these interviewees describe:

> Somebody didn't come back to me about a query that I made…they wanted to phone and I said 'well I'm sorry, I'm deaf'…and they never come back to me… it is hard [for hearing people] dealing with a deaf person I think, because they don't use the phone, it's difficult to communicate. I use Type Talk [a telephone relay service]…a previous friend, said 'oh it's so hard talking to you now [since she lost her hearing].' Yeah, [I thought] you should try it from this end [i.e., being deaf] [laughs]. It's frustrating, phenomenally so, and so tiring, trying to get hold of the information (Catherine, 57, recently deafened-hard of hearing, Internet user).

> …[when you are] out and about you know, shops or whatever then if people are talking…it can be quite difficult to communicate [edit]…in shops and stuff, you know…some people would be like laughing at you, or some people kind of avoid you…try and get somebody else to see you, because they think 'oh it's a deaf person' (Carol, 23, Deaf, Internet user).

Moreover, because D/deaf children find it difficult to communicate with hearing people they are commonly late in developing literacy skills. The different grammatical structure of sign language also means that some Deaf people find the grammar of spoken and written English problematic to learn. These difficulties can be further compounded by inappropriate or inadequate educational support at school. Until the mid 1980s the majority of D/deaf people in the UK were forced to study using oral methods of communication however ineffective these were in practice (exceptions were schools organized by the Royal National Institute for Deaf People). This refusal to allow D/deaf people the right to be educated in sign language meant that many D/deaf people lost educational opportunities. Policy has recently changed towards providing bi-lingual education, yet support in schools and universities for D/deaf people often remains inappropriate or unreliable. This, combined with the low expectations of some teachers, mean that D/deaf young people's levels of achievement remain below the national average and basic levels of literacy within the Deaf community are relatively low (Watson et al. 1999). As a result some D/deaf people find it difficult to access conventional sources of written information.

---

1   Rogers (1998) reports that when fires ravaged California in the 1970s hearing impaired people burnt to death because they were unable to hear public warning announcements.

I was quite frightened at the hearing school. I just didn't like it there, I was not happy. I wasn't allowed to sign, if I started to sign I was told off...I felt inside like so many frustrations and, and so angry about that and also I couldn't develop any English skills, you know I couldn't read, I didn't know how to talk...when I got to about eleven I moved to another school which had a deaf unit and a hearing unit but there used to be incredible clashes between the two. Lots of the hearing pupils used to tease the deaf pupils...when I was sixteen, I still had very poor English reading and writing skills (Alistair, 43, Deaf, Internet user).

Interpreters provide one potential means of mediating sign language users' access to information but are in short supply. Deaf people therefore commonly rely on hearing relatives or friends to access information or communicate on their behalf. In these situations Deaf people worry about the quality and depth of the information that may be communicated back to them, and whether this is comparable with the information that hearing people receive. Informal interpreters, such as relatives or friends can be unreliable. They may not translate the advice correctly because they are not properly qualified or they may appropriate an advocacy role, making decisions for the Deaf person, rather than merely acting as a conduit for information. Unlike professional interpreters, informal interpreters are not bound by a code of conduct and so may not respect the Deaf person's confidences. Indeed, Deaf people can be reluctant to use interpreters, professional or informal, in personal situations (e.g., for medical or financial matters).

For [job] interviews and things like that, I've told them I'm Deaf and my first language is British Sign Language and I need an interpreter and if there's not an interpreter I won't be able to go to interviews. Some people have said 'well, you know, it's going to cost too much money'...That type of discrimination I've faced...I've been to my GP and I've said look I need an interpreter and they don't know how to go about it (Jennifer, 26, Deaf, Internet user).

Prior to the development of the Internet D/deaf people accessed information from hearing society by using adaptive communication devices, in particular teletypewriter relay services which have different names in different countries (e.g., Type Talk in UK). This is a system where a D/deaf person can make or receive a telephone call using a hearing telephonist as a medium.[2] However, relay services are not popular with many D/deaf people because they are slow. Communication can only flow in one direction at a time so that the party receiving a message cannot interrupt or begin a response until the message has been completed. This can strip the communication of emotion, nuance and spontaneity. Given this form of communication also relies on written text some D/deaf people struggle with the

---

2   The D/deaf person types the information they wish to communicate, a hearing telephonist then calls the hearing person and verbalizes the information to them, and then in turn types the response back to the D/deaf person.

level of English required to use it, and because it requires an immediate response they do not have time to seek help in understanding the message or composing an appropriate reply. Echoing some of the concerns about using interpreters, D/deaf people also described being reluctant to engage in relay conversations where the information to be communicated relates to a personal matter. There are other practical limitations to relay services too: they rely on the D/deaf person being able to access a relay device, and the length of calls required for the communication can be expensive.

Moreover, some Deaf people are deterred from seeking information from mainstream organizations because of concerns that hearing advisers will not have the understanding of Deaf culture or the communication problems Deaf people encounter and so be unable to provide the appropriate information. Individuals also identified a fear of being 'judged' by hearing information providers if their spoken or written English is weak. Sign language users often prefer to seek information from Deaf organizations in their own language. However, a reliance on Deaf spaces for information presents its own particular difficulties. Deaf communities tend to consist of dense close-knit, locally-based social networks (Corker 1996). As such, some of the informants were concerned that their confidentiality may not be respected by Deaf organizations because gossip spreads quickly within any insular community.

Being excluded from the traditional oral-based information sources and communication possibilities has a profound impact on D/deaf people's knowledge of, and engagement with, mainstream society (Corker 1998). Studies show that D/deaf people as a group experience: economic disadvantage because they miss out on the knowledge from television/radio advertising required to make appropriate consumer choices (Schein 1989); have poorer health-care knowledge than hearing people (Woodroffe et al. 1998); and rarely vote in national elections because of their lack of information about the political process and consequently they experience civic disenfranchisment (Bateman 1998). The sense of injustice which flows from information poverty means that growing numbers of D/deaf people feel emotionally and socially detached from hearing society. At the same time the hearing world is ignorant of, and misinformed about, D/deaf people's lives (Rogers 1998). In this sense, traditional communication technologies are deeply implicated in how D/deaf people have been dis-abled in everyday life. The Internet however, has drastically changed D/deaf people's potential to access information without the need for it to be mediated by hearing people (directly as information providers or indirectly as interpreters or relay services) and to communicate remotely in sign language. For the first time Deaf people can simultaneously use sign language without co-presence of the signer and the sign-reader. The following sections explore how D/deaf people are using the Internet.

The Role of the Internet in Facilitating D/deaf and Hard of Hearing People's Access to Generic Information

Previous studies have suggested that marginalized groups including: disabled people and those from linguistic minorities have less access to ICT resources at home than the general population (e.g., Dobransky and Hargittai 2006). This gap has largely been attributed to income, or educational disadvantage. However, when the results from our project's survey of D/deaf and hard of hearing people was compared with findings from the General Household Survey (GHS) it showed that D/deaf respondents have levels of access to, and usage of, the Internet comparable with the general population. Indeed, 79 per cent of the D/deaf people who responded to our survey stated that they use the Internet everyday, compared with 59 per cent of those who completed the GHS. Given that the Internet has now become a mass medium (Willis and Tranter 2006) researchers have pointed out that questions about what people actually *do* with ICT are now more significant than questions of access. In particular, there is a need to understand: the role that the Internet plays in the everyday life of different social groups; how people develop relationships with it; and the connections between ICT and the capabilities people need to fully function in society (Holloway and Valentine 2003, Selwyn 2004). The evidence of this research is that the majority of D/deaf people who have regular access to the Internet engage with it in positive ways. Of those users responding to our survey 88 per cent said they found it easy or very easy to use. While traditional sources of information (particularly those available through libraries) require a significant level of English literacy, the simple language and visual content of most web pages, and the abbreviated English of email, provide a way of accessing information and a communication style that is more suited to D/deaf people. On-line communication allows D/deaf people to access information and to interact with hearing people to a degree that is rarely possible off-line because of their impairment and linguistic exclusion.

> I was brought up with a Deaf family so information was quite limited...now with the Internet you've got loads of information available to you [edit]...Before when you had no Internet you'd have to go to a library but I didn't really go very much because I can't be bothered to try and find books and try and read them. But since getting the Internet I've actually learnt quite a lot (Adam, 27, Deaf, Internet user).

As this quotation indicates the Internet is enabling D/deaf users to access a wide range of generic information for the first time which is broadening their understanding of, and ability to engage with, the hearing world. Barnett (2004) has noted that participation in discourses about matters of public importance consist of practices around the consumption of books, newspapers, radio, and TV programmes that are circulated by media and communications industries; while Kymlicka and Patten (2003: 13) have described democracy as a 'talk-centric process'. As such D/deaf people's off-line information poverty means they have been disconnected from public and political matters. However, on-line information practices now enable D/deaf people to engage more readily with current affairs and to participate in democratic processes (e.g., using email to hold political representatives accountable).

> I like to read the news [on-line]...you can find electronic magazines as well on the Internet, and I like to go to those magazines and newspapers and current affair things. See part of the problem...with being hearing impaired is that you don't know what's the hot topic at the moment, what people are talking about at the moment...because you don't hear it and you're out of it...So, by being able to go on the Internet, and learn things and find out things, I'm able to share those with my wife and with other people that I come into contact [with] (Boris, 55, Deaf, Internet user).

> I've found you know more information so I think it's you know broadened my mind a bit...you know, the Internet...the information there, different ideas...you can see different people's views (Raymond, 59, Deaf, Internet user).

> It's my umbilical cord to the world really...the Internet is, it truly is, I couldn't live with out it (Catherine, 57, recently deafened, Internet user).

More specifically, when our survey results were compared with findings from the GHS it showed that D/deaf respondents are significantly more likely to use the Internet than the general population to look for: health information (65 vs. 27 per cent), employment (36 vs. 24 per cent); and to email (95 vs. 80 per cent) or chat on-line (51 vs. 20 per cent). This use of the Internet for health information echoes findings from studies about disabled people's on-line information practices which have shown that disabled people experience an improvement in their quality of life because they are able to obtain a greater range and quality of medical information more quickly on-line than off-line (Dobransky and Hargittai 2006). For some D/deaf people on-line health information enables them to side step the confidentiality and communication issues that can occur when access to medical information is mediated though an interpreter. The depth of on-line information also gives them the knowledge to engage more effectively with health professionals, while email offers a new means of accessing help in medical emergencies.

> [I]f I'm feeling ill...I'll just go on to the Internet, and put in my symptoms and I'll bring up the possible things that could be wrong with me...Sometimes I'm not able to access the doctor and the information they give me fully, and so it's quite interesting that the Internet has that full information...and they've got a message board that gives up different people's views on specific illnesses (Jennifer, 26, Deaf, Internet user).

…[the Internet] it's actually kept me sane, through quite a difficult time of losing it [hearing]…to go on the web and find out information so that I was equipped to go and talk to consultants and say 'well what, what is that you're doing?' And 'why is that and why is this?' And like they put you on steroids, so I'd come back and I'd look it up [on-line]…found that I was on the maximum dosage, that I shouldn't have any more, that the side effects were this (Catherine, 57, recently deafened, Internet user).

D/deaf people also reported using the Internet to access goods and services. Over two thirds (69 per cent) of our survey respondents reported that they were good at searching for this information on-line. These findings are comparable with a similar survey in the US (n=227), where 76 per cent of deaf people who reported using the Internet said that they were comfortable or very comfortable doing so (Zazove et al. 2004).

Accessing information and communicating on-line enables D/deaf people to chose if, and when, they want to disclose their D/deaf identity in the same way that other studies have shown that disabled people manage information about their impairments on-line (e.g., Seymour and Lupton 2004). Of the respondents to our survey, 40 per cent of those who use the Internet reported that they conceal their D/deaf identity on-line, whereas others described using email as a tool to inform hearing people about how to communicate with deaf people. Both strategies potentially contribute to reducing the risk of D/deaf people being misinformed or discriminated against by hearing people. As a consequence D/deaf people are able to exercise their rights as both consumers and employees more effectively on-line than through face-to-face interactions with hearing people.

> Well, the positives about buying on the Net is that it is easy and you can shop around without any of the communication barriers of going into shops, looking in catalogues and phoning up for things which is a problem for Deaf people (Derrick, 34, Deaf, Internet user).

> I have to work with lots of companies for this job. A few years ago I had an experience where I mentioned to one company I was deaf and after that the company didn't want to know. So from there I am very careful about who I tell. It's a bit like when TypeTalk was set up, they [hearing run organisations] would identify that it was a deaf person [ringing] and would hang up but that doesn't happen on-line because it could be anyone (Samuel, 39, Deaf, Internet user).

There is some evidence that by communicating regularly with hearing people on-line Deaf people are gaining greater exposure to written text, and are therefore becoming more familiar with word patterns, a wider vocabulary and sophisticated grammatical structures that might enhance their literacy development. As studies with other social groups – particularly disadvantaged young people – have shown because using the Internet is regarded as enjoyable it can also be motivational. This interviewee explains:

I think it's better in terms of learning English, in terms of you know being more competent to write stuff...I've definitely learnt more, you know I've had to read more and stuff and see, you know. I think it's good in terms of you know Deaf people's exposure to English (Carol, 23, Deaf, Internet user).

Indeed, e-learning offers one potential strategy for (re)engaging D/deaf people who have had negative experiences of the education system in distance and life-long learning programmes. The Internet also offers the potential to support visual resources. The translation of more materials into sign language would increase Deaf people's access to a wider range of information as well as providing an opportunity for Deaf people to receive instruction and training in sign language and to interact with other learners and tutors in virtual signing classrooms and to receive signed feedback in real time. Currently, however this potential is largely unrealized because of the scarcity of on-line educational material in sign language; and the cost of the bandwidth necessary to support the multi-point continuous presence communication that is necessary for a sign language virtual classroom (Drigas et al. 2004).

In sum, in the first section of the chapter we have argued that because previous technologies such as the telephone are based on oral forms of communication D/deaf people have been limited in their ability to use such tools unless they are mediated by hearing people. As such D/deaf people have been excluded from information and communication possibilities with the consequence that they have not been able to participate readily in hearing society, nor exercise their citizenship rights. Although for some Deaf people their demand for formal recognition of BSL has stimulated a form of political participation and action (Skelton and Valentine 2003b). Indeed such demands for the recognition of BSL could have important ramifications for sign language provision on the Internet. However, despite the generic issues of marginalization and disconnection, this section has shown that the Internet is enabling D/deaf people to locate, use and communicate information *remotely* for the first time without the need for it to be mediated by hearing people. As such it is enabling D/deaf people to access many of the information and communication possibilities they were previously denied. This new degree of information literacy is enabling many D/deaf people to participate more effectively in the Information Society by giving them a greater ability to function independently, and therefore a perception of both self-confidence and equality with hearing people. Indeed, just under 70 per cent of the respondents to our survey said that using the Internet has improved their quality of life. These interviewees explain:

It's [the Internet] been brilliant, marvellous. Talking before the Internet you know we had to use minicoms or text, TypeTalk. I don't really like that, I mean it was difficult and you know it's expensive and so now you can email people everyday, its cheap, and easy to find out information...I don't have to bother my [hearing] wife to ask people for information on things and I don't have to get interpreters...so that's really good. I'm, it makes me more independent, I can find out things for myself (Raymond, 59, Deaf, Internet user).

I am a Deaf person, and I feel like, you know I'm equal to hearing people when I'm on the web, there's no difference (Carol, 23, Deaf, Internet user).

Hearing people can use the phones but deaf people can e-mail so it's more, more equal now...If I wanted to ask something...I needed an interpreter or TypeTalk...and I'd think, I can't be bothered. But now I can just e-mail if I want to know something...I can just do it myself it makes the Deaf community more confident. It means they're more equal to hearing people, they get access to the same information... (Adam, 27, Deaf, Internet user).

These quotations imply that the Internet may be facilitating D/deaf people's integration[3] in hearing society. Working with disabled people van de Ven et al. (2005) have defined integration in terms of five elements of interaction: functioning ordinarily without receiving special attention; taking part in society; trying to realize one's potential; directing one's own life; and mixing with others that are not disabled. This section has demonstrated that the Internet enables D/deaf people to achieve all five of these elements of interaction with hearing people to varying degrees. Thus, in van de Ven et al.'s terms, the Internet might indeed be considered to have facilitated D/deaf people's integration into hearing society. However, van de Ven et al.'s (2005) model of integration assumes that disabled people want to interact with, and be included in, existing mainstream society when this may not be the case. It also fails to acknowledge that non-disabled people may be inflexible, and unwilling, to accept the adaptations to mainstream 'norms' necessary to allow the inclusion of disabled people (Ryan 2006). As the evidence of this section of the chapter has demonstrated, while the Internet has undoubtedly been 'enabling' for D/deaf people, giving individuals more capability and therefore agency to function independently in hearing society, this has actually been achieved by allowing D/deaf people to 'pass' as hearing on-line and therefore to avoid having face-to-face contact with hearing people in off-line space. In other words, the Internet enables the incorporation of D/deaf people into hearing society – by allowing them to take up the subject position of hearing. Indeed, paradoxically the Internet is actually facilitating D/deaf people's existence in, but separatism from, the off-line hearing world (see also Valentine and Skelton 2008). As this interviewee explains: 'I don't think it's brought me closer to the hearing world, but I think it's enabled me to cope better in the hearing world.' (Boris, 55, deaf, Internet user).

Hence, the current structure of the Internet is contributing to the maintenance and normalization of hearing hegemony, without the necessity of hearing people having to make any accommodation for D/deaf people, thus leaving the

3   The concept of integration has been controversial because it is often associated with the notion of adapting to the norms and values of majority society.

discrimination D/deaf people encounter unchallenged. In the following section we reflect in more detail on role of the Internet in enabling Deaf people to communicate and exchange information with each other, rather than with hearing society.

Deaf Information Landscapes

Sign language is a visual, gestural and therefore deeply embodied form of language. Consequently face-to-face communication is at the essence of Deaf culture. However, the Internet is changing the way that these face-to-face interactions can take place because it supports visual forms of communication that, for the first time, make it possible for Deaf people to have synchronous, remote communication with each other in sign language. As a result there has been an explosion of information sharing between Deaf people and an increase in the density of Deaf people's social networks.

> It [the Internet] just makes it a lot more easier to be involved with Deaf people
> – because then we share more information so much easier and quicker...most of
> it can be done on the e-mail, because like sending pictures and that kind of thing
> and then you can do three-way communication kind of thing you know (Naudip,
> 37, Deaf, Internet user).

The Internet facilitates not just individualistic forms of communication but also community based communication. There is a burgeoning literature on the role of ICT in the creation of new forms of on-line community (e.g., Rheingold 1994). Over 85 per cent of the Internet users who responded to our survey said that they think the Internet supports the Deaf community. The Deaf community has always been a communication based community. Indeed, while 'the Deaf community is the first "community of relatedness" to emerge in the disability sphere' (Corker 1998: 135), it is not Deaf people's impairment but rather their shared language – sign language – that provides the powerful affective bond of belonging and collective political and social identity that binds the community together. Prior to the emergence of the Internet the Deaf community was based around a strong network of Deaf clubs, many originally called 'missions', established by Deaf people themselves (or through philanthropic institutions, often based on religion with a special concern for D/deaf people) in most major UK towns (Ladd 1998). The Internet however, has enabled Deaf Clubs to promote their activities and events beyond their own local networks and to reach out to those who are geographically isolated from Deaf venues. This has encouraged Deaf people to travel further afield to attend Deaf events – both nationally and internationally. Email (which is potentially a one-to-many form of contact and also offers advantages of convenience, and control in managing communication), and web cams that support remote signing, enable Deaf people to sustain social networks beyond their local geographical communities in ways that were not possible when sign language required physical co-presence and the telephone was the dominant form of remote communication.[4] Indeed, sign languages have an advantage over oral languages when it comes to international communication because there are strong grammatical similarities between the 200 sign languages of the world. Gestural languages are also more easily adapted to enable cross-lingual communication than oral languages. This process of blending signs on-line, alongside email, is facilitating Deaf people to initiate contact and develop friendships with people in other countries. In this way, the Internet has effectively enabled the Deaf community to be scaled-up from a local to a global network (Valentine and Skelton 2008) as these interviewees explain:

> I mean there's more information about the Deaf community on-line, there's information about Deaf clubs. I can find out information about different events and things are happening...there's a site called Deaf events which email me information (Raymond, 59, Deaf, Internet user).

> We've got web cams now and we can use this to communicate by sign and of course if that person's got a different language, we get used to using gesture... It's important for us to learn how to communicate like that. I've learnt a bit of international sign language like doing the alphabet on one hand and things like that...I've met people from France, Sri Lanka, Bangladesh, America, Australia, I'm trying to think, a variety of different sign languages involved there, but it hasn't been a problem, because most Deaf people can communicate via a Deaf [sic – web] cam (Jennifer, 26, Deaf, Internet user).

> I put in Deaf Club or Deaf Group in the Internet and all this information came up...Photographs of people, they were Deaf, said information about them...you know they're people from America...I couldn't have contacted anybody in America before the Internet (Deborah, 37, Deaf, Internet user).

The close-knit, localized nature of the material spaces of Deaf clubs, mean they can be experienced as quite 'closed' communities. Like all communities the Deaf community has its own cultural rules, which includes notions of belonging and those who do not belong. In particular, as a result of the oppression Deaf people have experienced in hearing society oralism has been discounted by the Deaf community in the same way that BSL is devalued by hearing society (Ladd 1996). There is a strong suspicion of 'hearing' ways: such as using the voice and within Deaf spaces those who ordinarily lip-read and speak in hearing

---

4   It is important to acknowledge however that mobile phones introduced similar possibilities of wider connectivity through SMS/text messaging. Young D/deaf people in particular made quick and extensive use of this medium to communicate. It was certainly an important element of recruitment of interviewees and facilitated greatly the arrangement of meetings for this research project.

environments 'switch off' their voice and sign. Hearing people in Deaf spaces are expected to sign rather than speak. Those deaf people who can 'pass' as hearing because of their oral communication skills, or who are strongly embedded in the hearing community and do not sign (e.g., some deafened or hard of hearing people), often struggle to be accepted as part of the Deaf community (Corker 1996). Nevertheless, the Internet has enabled these previously excluded groups of deaf people to access Deaf information and resources more readily, and to create their own specialist support networks (which often have a therapeutic role) and on-line communities. The ease with which different groups of D/deaf people can access D/deaf related information has also brought about changes in many of the Deaf clubs in the UK with many actively focusing on the 'wider' D/deaf community rather than just the BSL-using Deaf members who have traditionally been the mainstays of the clubs.

> I find out about any activity concerning hard of hearing, deaf people. For example, I was emailing a hard of hearing email discussion group...and occasionally they'll organise social evenings or days where we'll all travel to a certain place...meet up, it's brilliant for socialising, meeting new people (Samantha, 47, hard of hearing, Internet user).

Of course, Deaf people are not a homogenous group of Internet users. Our survey identified three distinctive groups of user: *Computer Savvy Users* (51 per cent of Internet users surveyed) – the majority of this group were BSL users and their average age was early thirties. This group use the Internet for work and leisure activities. In other words, the Internet was quite integrated into their everyday life. They also had a high degree of confidence in their own Internet skills. A second group is defined as *Tentative Users* (19 per cent of Internet users surveyed): the average age of this group was late forties and they were relatively new to computing. They reported that they mainly used the Internet at home and tend to be quite fearful about using it beyond very standard activities. The third group, *Instrumental Users* (30 per cent of Internet users surveyed), had the highest level of educational qualifications and were more likely to have a postgraduate degree than other users. Their use of the Internet can be described as 'business-like' because they mainly use it for activities such as banking and work rather than for leisure purposes. Unlike *computer savvy users*, this group's use of the Internet was not integrated into their wider everyday lives. In addition, the survey identified a further group who do not access the Internet at all. For this group barriers to the Internet not only included cost and lack of ICT skills, but also low levels of literacy which limit their ability to access information in written English. This does not mean however that Deaf non-users and *tentative users* are excluded from some of the benefits of Internet use. Rather, key individuals within local off-line Deaf communities can play an important part in disseminating information from the Internet to wider groups of Deaf people who have limited access to, or ability to use ICT, or who chose not to use the resources at their disposal. These key individuals also contribute to the dissemination of ICT skills within Deaf communities and provide technical support to other Deaf users – both formally through some Deaf clubs, and informally through their own social networks. Such opportunities, to draw on local social networks for technological expertise and support, have been identified as a critical factor in promoting sustained ICT use (Murdock et al. 1996). This Internet user describes how she contributes to the wider transmission of information and skills amongst Deaf people.

> It's been five years since I've been using the Internet...and I'm providing a service to Deaf clients now. Part of my job is really to teach them how to use the Internet to search for jobs or you know fill in on-line application forms... [later she continued] I'm able to extract information myself, read through it and put it into BSL for BSL users... Also if a person comes to see me and say they've got a problem with say their benefits and, or it's too complicated, they can't read the information well, I have a look at the on-line information and then translate it for them (Jennifer, 26, Deaf, Internet user).

Of course, as this quotation hints, there are also limitations to the ways that the Internet can contribute to enabling Deaf people to be included in hearing society and there are differing levels of engagement with this technology within the Deaf community. Despite the possibilities for information to be provided in sign language clips on-line and the use of plain English on many websites, nonetheless the standard of English on some Internet sites can exclude Deaf people whose levels of English literacy are limited. Rudd (2002) has talked about 'content chasms' being a key digital divide issue and the need to develop on-line content created by/for socially excluded groups, this is certainly pertinent in relation to sign language users. Ironically, despite the UK Government's commitment to promoting on-line participation its websites were identified by interviewees as some of the worst culprits in terms of the density of the written English and the lack of Deaf awareness evident in their content and visual presentation.

> I mean in some sites the English is kind of really high level, so if it's the Government websites or stuff like that then the English can be quite complex. And if it's that then I kind of just ignore it and don't bother to read it. It would be really good to maybe have an interpreter involved or you know some BSL, so it's easier for a Deaf person just to understand the information that's on the web (Matthew, 22, Deaf, Internet user).

Indeed, Castells (2001) suggests that soon the most problematic dimension of the 'information rich'/'information poor' digital divide will be the 'knowledge gap' caused by constraints on individuals' capacities to make effective *use* of digital resources rather than inequities in *access* to technology. In this context, information literacy – in addition to, and distinct from, IT literacy – is a far-reaching educational and democratic issue in the information society. It was evident from this research that some Deaf people in particular lack essential information literacies to enable them to make effective use of on-line resources. This is related not only to language issues but also to a general naivety about information sources because of D/deaf people's off-line information poverty. This interviewee explains:

> When I had a look on the Internet for information there was so much information there so I was like wow this is fantastic, so I just believed everything. I thought wow that's all true, that's fantastic...but then I realised that you know there's some false information...then I realized...don't believe everything you read (Jacob, 27, Deaf, Internet user).

The misunderstanding of on-line information by individuals can also be particularly damaging within the Deaf world because the density of Deaf social networks (described above) mean that misinformation can be widely transmitted.

In sum, the Internet is providing positive opportunities for D/deaf people to access the Deaf world and associated support, and for Deaf communities to be scaled up from a local to a global network. As such the Internet is enabling individual Deaf people to support and sustain their self-identities as Deaf, and it is enabling some individual deaf people to find out about Deaf culture and to possibly take up the subject position 'Deaf' for the first time. In such ways, the Internet is enabling the promotion and development of Deaf culture within and beyond the nation state. The benefits to D/deaf people and Deaf communities are not, however, evenly accessible because some Deaf people, like other populations too, lack the English, and information literacies necessary to participate in on-line worlds. As such, there is a need for more support for those who have least access to both technology itself, and the associated skills necessary to be able to take advantage of it. The conclusion reflects further on the complexities and paradoxes of D/deaf people's Internet use.

Conclusion

In this chapter we have demonstrated how being excluded from traditional information sources and communication possibilities has had a profound impact on D/deaf people's knowledge and awareness of: mainstream society; educational and economic prospects; social welfare and well-being; and political participation and civic engagement. In this sense, traditional communication technologies have been deeply implicated in how people with hearing impairments/sign language users have been dis-abled in everyday life. The emergence of the Internet, however, has opened up the potential for D/deaf people to access information without it being mediated by hearing people (directly as information providers or indirectly as interpreters/relay services) and to have synchronous, remote communication in sign language for the first time.

The chapter detailed how D/deaf people are using the Internet, firstly in the context of hearing society, and secondly, to exchange information and communicate with each other. In doing so, we have shown that D/deaf people's on-line practices are drastically improving their access to general and Deaf-related information resources, thereby enhancing their personal information landscapes. In this sense, the Internet has been enabling for people with hearing impairments and sign language users because it has given them new capabilities to function independently in hearing society. Indeed, D/deaf people mainly use the Internet in capital enhancing ways (e.g., to access entertainment, learning, consumption and employment opportunities; and to develop their social networks). As such, the benefits they gain from the Internet appear to be more positive than those realized by people with other impairments. For example, Dobransky and Hargittai (2006) claim that disabled people who participated in their study gained only limited benefits from the Internet because they mainly used it for a narrow range of purposes: to play games and to access health information.

However, the Internet, although enabling people with hearing impairments and sign language users to participate more effectively and independently in hearing society, does not result in the integration of D/deaf people into the hearing world. This is because hearing people are not being forced to change their own communication practices to accommodate D/deaf people's needs. True integration is a two-way process, it requires not just that D/deaf people adopt hearing ways of accessing information and communicating on-line but also that hearing people must adapt their on-line practices to facilitate D/deaf people's needs. Yet, to date many of the benefits D/deaf people gain from the Internet accrue precisely because it enables them to effectively take up a subject position as hearing: 'passing' on-line. Instead, of producing new normativities that integrate D/deaf and hearing people's communication styles, most mainstream websites actually reproduce hearing hegemony through their emphasis on written text, and increasingly also, audio-content or multi-media formats which provide a growing threat to D/deaf people's access to mainstream on-line information. Moreover, the Internet, by facilitating Deaf people to access entertainment, consumption, and learning opportunities without engaging face-to-face with hearing people is effectively weakening contact between Deaf and hearing people in the off-line world. At the same time the Internet is strengthening and internationalising Deaf networks and hence contributing to Deaf separatism off-line and on-line. Consequently, the Internet is not producing either just positive or just negative outcomes for D/deaf people but rather is generating a complex set of paradoxical effects for different users.

If the Internet is to contribute to the integration of D/deaf people it needs to accommodate the communication styles and needs of *both* sign language users and those with hearing impairments who communicate orally. A recent Disability Rights Commission Report (2004) claimed that the provision of accessible web content is often an afterthought in the process of website design/implementation. It argued that achieving full accessibility requires designers and implementers to incorporate different users' needs as a pre-requisite from the earliest stages of development, and that websites should be tested with a panel of users across a range of abilities and preferences.

For Deaf people whose first, or preferred language, is sign language and who commonly have limited literacy levels, their integration into the information society will only be truly 'meaningful'[5] when they are able to access on-line content in their own visual language on general websites. To achieve this more on-line content needs to be created by, and for, this socially excluded group. Sign language can be delivered on-line via video clips of human signers[6] (computer generated signing animation is a long way from widespread commercial application) (RNID 2004). For deaf and hard of hearing people audio-visual clips on web pages need to be subtitled and technical support needs to be provided through communication media other than telephone help-lines. It is important that on-line content for both Deaf and deaf people is fully equivalent to that provided in other presentational forms (e.g., text or audio content), not a simplified substitute that contains summarized information.

However, D/deaf and hard of hearing people have relatively limited economic power to compel web designers to meet their needs and mainstream hearing organizations have limited incentives to change their practices to accommodate minority groups of users. As such, there are potentially strategic benefits for D/deaf people in forming coalitions with other disabled and minority language users to lobby for regulation to enforce on-line content to incorporate a greater range of users' needs. However, even governments have a limited ability to regulate the Internet compared to the telecommunications industry (in the past public ownership of the broadcast spectrum allowed the UK Government to mandate the industry to provide services for people with hearing impairments) (Rogers 1998). While the UK *Disability Discrimination Act* requires organizations to make reasonable accommodation for people with impairments – which should include accessible websites – its explicit application to on-line content has yet to be tested in the law courts. Even if governments can require organizations to make commercial and consumer information accessible for Deaf people, Rogers (1998) argues that 'it is impossible to impose the same requirements on millions of individual computer users who effectively become broadcasters and publishers when they go on line without compromising free speech'. Notwithstanding this complication, if governments are truly committed to improving the possibilities for all citizens to participate fully on equal terms in an information society they need to accommodate the needs of different groups on their *own* websites and in terms of support for training and access. For D/deaf people this entails: providing information in accessible formats; funding or supporting voluntary organizations

---

5   Selwyn (2004) argues that Internet use is not necessarily 'meaningful' unless the user has control and choice over the technology and its contents. It is this that produces meaning, significance and utility for individual users.

6   They are however, expensive to make and hard to amend which means in a space like the web where information is fluid these clips can be quickly outdated.

to improve D/deaf people's access to on-line information; subsidising on-line subscriptions for D/deaf people on low incomes and supporting the development of affordable video compression and transmission technology for video communication using sign language.

While there is no necessary connection between integrating disabled and minority language users' on-line and changing attitudes to these groups off-line, we can theorize that because on-line and off-line worlds are mutually constituted (Valentine and Holloway 2003), changes in on-line information and communication practices might contribute to producing changes in off-line communication conventions and attitudes towards minority groups. For example, more on-line content in sign language might produce greater awareness amongst hearing people of D/deaf people's communication needs and preferences, and enable oral communicators to pick up the basics of sign language, which in turn might translate into breaking down some off-line hearing normativities and challenging oralist conventions embodied in civic citizenship.

Acknowledgements

We wish to acknowledge the support of the AHRC that funded the research on which this chapter is based.

References

Barnett, C. 2004. Media, democracy and representation: disembodying the public, in *Spaces of Democracy*, edited by C. Barnett and M. Low. London: Sage, 185-296.

Bateman, G.C. 1998. Deaf community and political activism, in *Cultural and Language Diversity and the Deaf Experience*, edited by I. Parasnis. Cambridge: Cambridge University Press, 146-159.

Castells, M. 2001. *The Internet Galaxy: reflections on the Internet, Business and Society.* Oxford: Oxford University Press.

Corker, M. 1996. *Deaf Transitions*. London: Jessica Kingsley Publishers.

Corker, M. 1998. *Deaf and Disabled or Deafness Disabled*. Buckingham: Open University Press.

Disability Rights Commission. 2004. *The Web: Access and Inclusion for Disabled People*. The Stationery Office, UK. Available: http://www.tso.co.uk/ bookshop.

Dobransky, K. and Hargittai, E. 2006. The disability divide in Internet access and use. *Information, Communication and Society*, 9(3), 313-334.

Drigas, A.S., Vrettaros, J. and Kouremenos, D. 2004. E-learning environment for Deaf people in the e-commerce and new technologies sector. *WSEAS Transactions on Information Science and Applications*, 5(1), 24-31.

European Council. 1994. *Europe and the Global Information Society* (Bangemann Report), European Commission. Available at http://www.rewi.hu-berlin.de/jura/proj/dsi/report.html [accessed 9 August 2002].

Foley, P. 2004. Does the Internet help to overcome social exclusion? *Electronic Journal of e-Government*, 2(2), 139-146.

Gleeson, B. 1999. *Geographies of Disability*. London: Routledge.

Holloway, S.L. and Valentine, G. 2003. *Cyberkids: Children in the Information Age*. London: Routledge Farmer.

Kyle, J. and Woll, B. 1983. *Language in Sign*. London: Croom Helm.

Kymlicka, W. and Patten, A. 2003. Language rights and political theory. *Annual Review of Applied Linguistics*, 23, 3-21.

Ladd, P. 1988. The modern deaf community, in *British Sign Language*, edited by D. Miles. London: BBC Books, 27-43.

Ladd, P. 2003. *Understanding Deaf Culture: In Search of Deafhood*. Clevedon: Multilingual matters.

Lane, H. 1997. Construction of deafness, in *The Disability Studies Reader*, edited by L.J. Davis. London: Routledge, 153-171.

Miles, D. 1988. *British Sign Language*. London: BBC Books.

Moser, I. 2006. Disability and the promises of technology: technology, subjectivity and embodiment within an order of the normal. *Information, Communication and Society*, 9(3), 373-395.

Murdock, G., Hartmann, P. and Gray, P. 1996. Conceptualising home computing: resources and practices, in *Information, Technology and Society*, edited by N. Heap, R. Thomas, G. Einon, R. Mason and H. Mackay. London: Sage, 269-83.

Oliver, M. 1984. The politics of disability, *Critical Social Policy*, 4(2), 21-32.

Rheingold, H. 1994. *The Virtual Community: Finding Connection in a Computerised World*. London: Secker and Walbury.

RNID. 2004. *Deaf and Hard of Hearing Users and Web Accessibility*. London: RNID Report.

Rogers, T. 1998. Access to information on computer networks by the Deaf. *The Communication Review*, 2(4), 497-521.

Rudd, T. 2002. Looking into the digital divide(s) and seeing a content chasm: a discussion paper. Coventry: British Educational Communications and Technology Agency (BECTA).

Ryan, S. 2006. It takes two to tango...but what if one can't dance and the other doesn't want to: a response to van de Ven et al. *Disability and Society*, 21(1), 91-92.

Schein, J.D. 1989. *At Home Among Strangers*. Washington DC: Gallaudet University Press.

Selwyn, N. 2004. Reconsidering political and popular understandings of the digital divide. *New Media and Society*, 6, 341-362.

Seymour, W. 2005. ICTs and disability: exploring the human dimensions of technological engagement. *Technology and Disability*, 17, 195-204.

Skelton, T. and Valentine, G. 2003a. 'It feels like being Deaf is normal': an exploration into the complexities of defining D/deafness and young D/deaf people's identities. *The Canadian Geographer*, 47(4), 451-466.

Skelton, T. and Valentine, G. 2003b. Political participation, political action and political identities: young D/deaf people's perspectives. *Space and Polity*, 72, 117-134.

Valentine, G. and Skelton, T. 2007. The right to be heard: citizenship and language. *Political Geography*, 26, 121-140.

Valentine, G. and Skelton, T. 2008. Changing spaces: the role of the Internet in shaping Deaf geographies. *Social and Cultural Geography*, 9(5), 469-485.

van de Ven, L., Post, M., de Witte, L. and van den Heuvel, W. 2005. It takes two to tango: the integration of people with disabilities into society. *Disability and Society*, 20(3), 311-329.

Watson. L., Gregory, S. and Powers, S. 1999. *Deaf and Hearing Impaired Pupils in Mainstream Schools*. London: David Fulton.

Willis, S. and Tranter, B. 2006. Beyond the digital divide. *Journal of Sociology*, 42, 43-59.

Woodroffe, T., Gorenflo, D., Meador, H. and Zazove, P. 1998. Knowledge and attitudes about AIDS among deaf and hard of hearing persons. *AIDS Care*, 10, 377-386.

Zazove, P., Meador, H., Derry, H., Gorenflo, D., Burdick, S.W. and Saunders, E.W. 2004. Deaf persons and computer use. *American Annals of the Deaf*, 148(5), 376-384.

Chapter 6

The Geographies of Interdependence in the Lives of People with Intellectual Disabilities

Andrew Power

Introduction

People with intellectual disabilities (PwID) are a unique group, largely at the periphery of geographic enquiry and arguably left absent from the social model of disability. The lasting confusion and multiplicity regarding the definition of PwID has somewhat led them to being placed off the map with regard to their place in disability studies, as well as requiring the need for a more nuanced account of their lives. Intellectual disability (ID) is a problematic and complex concept which is difficult to define. Having an ID cannot be defined as a biological attribute per se because the meaning and interpretation of ID are negotiable (Ryan 2008) and vary across time and place. Generally they are defined as having long term impaired intelligence and social functioning (Department of Health 2001). They have been described as being particularly vulnerable to dependency creation on the one hand (Swain, French and Cameron 2003), yet, can be highly independent on the other (Lemon and Lemon 2003). These almost unique characteristics have led to much debate about the role of carers in the lives of PwID and their 'place' in this quandary. This has significant theoretical and policy concerns for researchers. This chapter attempts to develop a geography of ID in light of this key theme.

The historical development in the thinking regarding the definition, treatment and advocacy of ID reveals a deep-set ethos, which assumed that others were responsible for looking after PwID, and that little could be expected of them. Over time, it has been possible to see a change in this ethos, where PwID have proven over and over that they are capable of far more independence than many had expected, and policy makers, researchers and 'advocates' gain a vision of greater changes that are possible. However, the self-advocacy movement for people with intellectual disabilities lags far behind many other civil rights efforts, such as those related to physical disabilities. Furthermore, the assumed dependency of PwID has left behind a legacy in terms of attitudes regarding their 'care'.

Research into the geographical experiences of PwID has emerged quite recently and has been described as 'at best, a footnote to debates on post-asylum geographies' (Hall and Kearns 2001: 238). However, in a short while, some studies of note have been able to unpack the complexity of the geographies of this group.

This chapter will draw out the geographical contribution to the complex theme of interdependence by tracing the socio-spatial lives of PwID in the community.

Today, most PwID are cared for at home where parents continue to grapple with issues of independence for their child. At the same time, for people living in traditional institutions and group homes, service providers are beginning to share the same goal as families in trying to create more independent living arrangements for people in their care. Both families and service providers are therefore trying, to varying extents, to ally various sources of support in their community. The main focus shall be on PwID living at home as it reveals the unique lifeworlds of young adults with ID grappling with independence in their community. It draws on important insights from disability studies and family care literature and then explores the geographical contribution to these debates.

The central aim is thus to examine the complexity of the many interconnections and interdependences across space and scale between PwID and the people that exist in their caring nexus such as family carers, siblings, formal care workers, social advocates, and so on. It is the relations, attitudes, and behaviours of these actors within this nexus that shape an individual's experience of place and well-being. The chapter plans to develop a social ecology model as a way of unpacking the fluidity and complexity of linkages and interactions involved in this caring nexus across different places. Places are the point at which these communities and networks come together. Often for people with ID, these are still in segregated settings in the community. As Laws and Radford (1998) state, because of their social networks being 'outside the mainstream', PwID have a complex geography.

The chapter starts by examining the issues of care and dependency for people with ID. It identifies, firstly within the family, the spatially contingent nature of caregiving relationships. The second section broadens the scope of care relationships by examining the wider nested dependencies inherent in the lives of PwID in the community. With the delicate balance between care and dependency in their wider locality, the third section examines the possibilities and barriers across space and scale to developing an interdependent life. Throughout the chapter, the discussion takes a spatially informed approach to care that looks at the multiplicity of care relationships as well as the socio-spatial contexts in which they occur. By examining their socio-relational geographies, the chapter provides a more nuanced reading of the geographical lifeworlds of PwID.

Care, Dependency and the Geography of Care

Over the last 15 years, family carers have become more noticed and are regarded as central to the care of PwID, particularly children and young adults living in the community. This section examines the central issues of care and dependency involved in these care relationships. While caregiving research has had a long history, it began to flourish from the early 1990s due to the resettlement strategies of deinstitutionalization. This literature generally focuses on the carer and cared for dyad, which tends to pose the scenario where the birth of a child with an ID often has a profound effect on the family, requiring one or more parent adopt the role of 'carer'.

Traditionally, this research has highlighted the stresses or 'burden' imposed on parents by the many, and often complex, demands of caregiving (Hassall et al. 2005, Shearn and Todd 2000). Birenbaum (1971) suggested that family coping becomes more difficult as the child with ID ages and as families face support network shrinkage over time. Similarly, there has been a focus on the health-related stress and depressive symptoms experienced by caregiving mothers of adults with ID (Pruchno and Meeks 2004). Work has also been done on coping strategies of ageing mothers of adults with ID showing that people adjust to caregiving over time, and acquire skills and competencies which help them to cope better, even when their own support networks may indeed be more depleted (Seltzer and Heller 1997).

Meanwhile, work within the geography of care has largely ignored the care of PwID, concentrating on the care of the elderly, or people with a physical disability or mental illness. At the broad experiential level, this literature identifies many trends that family carers typically share across a range of different care situations including that of PwID. Milligan (2001), for example, identified that with the increased use of the home as a carespace, many carers feel there has been an institutionalization of the home. Similarly, Wiles (2003) found that carers' mobility and routine is greatly affected having implications for their experiences of the home and local community.

The concept of caring in the disability field has come under much more scrutiny than in the elderly literature however. This is particularly relevant to the care of PwID because of the way in which care has historically been defined. In reaction to much of the caregiving literature (and to the rise of carers as a lobby group more generally), there has been much antipathy to the concept of 'care' in disability studies. In essence, the care system is viewed as central to disablist structures and practices in society (Thomas 2007). With the long history of paternalistic and professional intervention, those involved in caring are held responsible for creating and exacerbating dependency by denying PwID the means by which to govern their own lives. Moreover, it has been argued that 'family care' is *not* exempt from dependency creation; it neither guarantees greater independence nor autonomy for disabled people (Morris 1993). Often parents are seen as part of the problem. Authors such as Swain, French and Cameron (2003) argue that PwID may be particularly over controlled by their parents and carers and denied opportunities for experimentation and choice. At the same time, PwID may find it difficult to initiate change and parents may be reluctant to do so because of upheaval, risk and threat to existing sources of support.

Geographers examining the care of PwID must be aware therefore of possible conflicts that the concept of care can present. On the other hand, geographers have the potential to unpack and examine these themes – by examining the way these conflicts play out across space and scale. One example is Power's (2008) study that offers a more nuanced account of the conflicts family carers face in trying to promote independence. His findings reveal that those who provide assistance can often be 'caught in the middle' between trying to provide that care and simultaneously risk overprotection across different sites in the community. By focusing on the socio-spatial 'fine print' (Hall 2005) of the everyday lives of PwID, it is therefore possible to start to unpack how the issues of independence, advocacy and dependency play out across space.

A critical approach to the geography of care must be adopted therefore, where conventional approaches to carers are challenged and the notion of care is reframed as an interdependent and connected concept (Phillips 2007). Locating families within networks is a fruitful way of looking at their contribution to care vis-à-vis other actors. It is often these other actors that get ignored in the geography of care literature. Exploring social networks outside the typical carer/cared-for dyad, by means of a social ecology approach is a useful tool, particularly for geographers, to investigate the complex daily lives of people with PwID. This stresses the importance of the relationships that PwID have with others.

In particular, public space, where stigma can still be a very real issue for PwID, can have dramatic effects on care and dependency of this group. Public places, according to Valentine (1996: 216), do not just exist but are actively produced through repeated performances and are subject to the effects of power and exclusion. Parr (1997) argues that people must face 'sane-itising' regulations and understandings of how the self should be presented in everyday life. Therefore, as Ryan (2008) found, public interaction has specific consequences for mothers taking their children with ID out, as the actions of the child, the mother and others present make visible the rules of what is considered to be acceptable behaviour and etiquette in public places. Children with ID may demonstrate unconventional behaviour within public places, such as making loud noises, jumping up and down on the spot, flapping their arms or touching others present (e.g., Gray 2002). Although encounters in public places are often fleeting, for the family member or PwID there is a repetitiveness of these 'fleeting' moments, as they happen again and again. For families, therefore, going out in public space can involve considerable layers of negotiation, mediation and management (Read 2000).

In Ryan's empirical study (2008), in the dynamic interactions between carer, cared for and the public, she found that parents sometimes apologized, sometimes made excuses, or else ignored public reactions entirely to the behaviour of their child with ID. In all these cases however, they were all regarded as being obliged to assume the management control of stigma and the impairment effects associated with an ID. As Power (2008) found, in this more nuanced account of the carer/cared for negotiation, carers can often be challenged by constructions of independent living and having to promote independence, particularly in public space. In this complex geography, coping with stigma can have profound effects on carers' ability to promote independence on the one hand, and grapple with overprotection on the other. Carers often have a strong role to play in how PwID achieve independent living, defined as achieving control, choice and self-governance in one's life

(Morris 1993). In particular, those with Down's Syndrome can be imbued with having a need for care, despite often being highly independent. According to one parent of an adult with Down's Syndrome, in Power's (2005) study:

> A lot of the time, the people you see walking up the town, taking [an adult] with Down's Syndrome [DS] by the hand; it's almost accentuating their disability. I think they're a very empathic people – people with DS you know, they really live up or down to your expectations. And if you believe that he or she would never be capable of something, then a lot of the time, that's what happens (Carer R.21; source Power 2005).

Clearly then, the geography of care can identify the role that public space can play in the way care and dependency is shaped. To complete this picture, geographers must expand their analysis to include other actors which exist within the caring nexus. In most of these caregiving debates, there is a dichotomization of the 'carer' and 'cared for'. This dyad promotes a notion of the lack of agency and reinforces the paternalistic form of care with a passive dependence of one person on another, as well as ignoring the wider important relationships. According to Phillips (2007) one of the key shifts needed in a reconceptualization of care is a blurring of the boundaries of care and not to make false dichotomies of 'carer' and 'cared for' if the concept is to be valued. By blurring the boundaries we gain a new perspective on care at the margins and the intersections between private and public, professional and personal, and paid and unpaid. Care involves reciprocity and interdependency. It is part of a wider set of network relationships.

Even within family dynamics, the typical carer–cared for dyad oversimplifies the care relationship at home. The micro-geographies of the home space reveal that, as Phillips (2007) states, the typical 'caring' family no longer predominates – there are many families with many configurations of care. The complexity of relationships within families makes it impossible to be categorical about who depends on whom. Furthermore, research which construes the home and family as a private domain, only to be intruded upon by the state in times of emergency or dysfunction, has prevented the complexity of the ordinary everyday organization of caring tasks and activities that go on in the home space to inform research in areas which relate to care of PwID.

Similarly, outside the home, there are many important actors in the lifeworld of PwID such as siblings, extended family, members of the public, advocacy agencies and mainstream local service providers. It is therefore important to also explicate the relationships a PwID makes in various places and the basis of power in these relationships. The context of these relationships is important, as care has for so long been associated with custodial, paternal, protective forms of care, rather than developmental.

Kittay's (1999) notion of 'nested dependencies', as a fundamental part of the lifecourse, is helpful in this approach, in identifying the range of actors involved in caring relationships. Nested dependencies link those who need support with those who help them and which in turn link the helpers to a set of broader supports. This has significant theoretical and policy concerns for geographers, as it opens the focus of analysis up to all the linkages made across the whole nexus of care. A small geographical literature is developing that explores these relationships between disabled people and their family caregivers, formal service providers, bus drivers and others who contribute to creating their social space. Crooks, Dorn and Wilton (2008) argue that these social networks are integral to creating or eliminating barriers for those with various impairments. The following section examines this concept in more detail and begins to unpack the complex geography of PwID across these wider relationships in their lives.

Nested Dependencies

Geographers in a relatively short space of time have been able to capture the complexity of the socio-spatial lives of PwID. Social policies for PwID are generally centred on their 'reinclusion' into mainstream socio-spaces through engagement in 'normal' activities, primarily paid employment and independent living. However, deinstitutionalization, whilst bringing PwID physically into society has left them, in many cases, socially isolated and largely invisible (Hall and Kearns 2001). Geographers have begun to identify where these nested dependencies play out, offering a more nuanced account of the spaces in which PwID live.

Immediately outside the carer–cared for dyad, it is possible to see the caring role other family members play and the effects on them. According to a study by Mulroy et al. (2008), many parents reported the ease with which siblings went to the aid of their parents in caring for the child. In addition, this was generalized to situations outside the family home, with sentiments like: '[the child was] very helpful especially at school to those who are a little slower and need a bit of extra help' (225). Furthermore there were many positive benefits reported of the siblings such as increased tolerance, being acceptant of difference and being more mature than similar aged peers. Another common advantage as stated by most parents in the study was that their other children were notably more compassionate, caring and patient both in a general sense as well as in their role as 'teachers' to their siblings (Mulroy et al. 2008). In essence, this sharpening of values of the other sibling(s) has mutual and profound effects on the life of the person with ID. A relational geography thus already shows that examining the interactions within the broader caring nexus can change the experience of space in both the home and beyond.

Outside of the home, PwID can appear to have very similar geographies to non-disabled people on the surface. They may live with their family or in a private home and may use public transport to go to work (Laws and Radford 1998). However, this represents only one end of a spectrum of ID; the vast infrastructure of 'special' schools, 'special' transport and sheltered workshops still comprise the daily geography for many. Furthermore, when the differences in the social activities in public places are considered, often the geography of a PwID can represent quite a routine use of a city's resources (Laws and Radford 1998).

A closer inspection of their everyday social geographies often reveals low levels of interaction with non-disabled people, small action spaces, and precarious finances, all of which point to a 'life on the outer fringes of the daily round' (Laws and Radford 1998). By taking a social ecology approach, it is possible to identify the many networks and linkages made between the PwID and family members, care managers, speech therapists, home-helps, social workers, drama and arts teachers, sheltered workshop staff, as well as members of staff, bus drivers, and benevolent members of the public. First and foremost geographers must understand the relationships that exist in the places they use. It is these relations, attitudes, and behaviours that shape an individual's experience of place, and without a doubt this is central to well-being. Places are the point at which these communities and networks come together. When choices are available, people often choose to develop their identities in accordance with particular places. PwID however rarely have those choices available to them (Laws and Radford 1998). Their unique position of being overly susceptible to dependency can have profound impacts on their geography; it is in the particular everyday places that PwID interact with other people.

Some studies on the social ecology of PwID reveal the characteristics and extent of these personal networks. Hatzidimitriadou and Forrester-Jones (2002) reported that the social networks of older PwID (who had lived with their parents or in community settings all their lives) included 32 per cent other service users, 12 per cent care staff, and 11 per cent family members and also a relatively high proportion (30 per cent) of community members (neighbours, other friends and contacts from mainstream clubs, e.g., Age Concern). For individuals living in community residences and engaging in supported employment, Forrester-Jones et al. (2004) reported a mean network size of 36 members for 18 young adults. In another study, Forrester-Jones et al. (2006) found the average network size for adults (average age 53) living in the community 12 years after resettlement from long-stay hospitals, was 22 members with a range of 3-51. Just over a third of the total network members were acquired from community contexts, separate from ID services including clubs, church and voluntary organizations (7 per cent), retail services such as shops, pubs, cafes, cinemas (6 per cent), neighbourhoods (4 per cent), family (14 per cent) and other friends and acquaintances (10 per cent). However, the same study found that staff, including cleaners and other support staff, were still the main providers of most types of social support and most likely to be in contact with respondents on a daily/weekly basis.

To understand this social relational geography, Hall (2005) urges us to move away from assuming that marginalization equates solely with exclusion from mainstream social activities and spaces, and constructing inclusion as a process of incorporation into these activities and spaces. This is particularly the case if PwID face a mix of patronization, fear, an unwillingness to understand 'non-standard' forms of communication, and a strong sense of 'difference' in public life (Hall and Kearns 2001). This is a significant step in the geographical literature on the deinstitutionalization of PwID, from being dominated by the binaries of institution/community, isolation/integration and individual/social (Gleeson and Kearns 2001). Forrester-Jones (2006) found that the type of accommodation in which people were living had a strongly significant effect on the types of social support which they received. Unfortunately, low scores for personal support were associated with living in hostels, small group homes and especially supported accommodation. However, when considering material support, the situation is reversed: here living in hostels, small group homes and supported accommodation is associated with higher scores for material support; people living in residential and nursing homes reported lower levels of such support. Similarly, Hall (2005) identified participants who, having secured flats of their own, now feel more dependent on support staff, who are continuously present. Geographers must be critical of the spaces of incorporation and not assume a place of incorporation is the best choice for PwID. Often the social relations at work in the sheltered workshop, the group home, or the boarding house are *not* the ones that promote autonomy and self-esteem (Laws and Radford 1998).

Essentially, for those living in the community, geographers must be concerned about whether their nested dependencies are becoming more fragmented or cohesive. A focus on nested dependencies forces us to think beyond the mainstream/ segregated binary to envisage different uses of the same sites. For instance, recent policy statements about services for adults with intellectual disabilities in the UK have pushed for a reoriented day services model based on a 'dispersed' or 'centre-less' service. This means that services are geared towards accessing 'mainstream' community amenities and facilities, rather than scheduled attendance at special day services centres (Simpson 2007). These may include garden centres, coffee shops, bowling centres and so on. These can prove popular for those with milder forms of ID; yet remain uncertain for those with more complex needs. This would suggest the decline of the symbolic role day services centres play as a physical and fiscal commitment to public service provision. In other words, PwID may see ever more dispersed nested dependencies across the community in future, particularly with reports such as 'Valuing People Now' advocating that day centres need to close (2008). Proposing a more fluid and reflexive relationship between public and private space allows for a more nuanced account of where care takes place.

It is not enough to assume that once PwID are 'normalized' into the social roles of a non-disabled person in a dispersed setting, they will experience less discrimination and be incorporated into other mainstream social spaces and activities. However, through cooperative structures and community entrepreneurial activity – with an emphasis on collectivization and care – many PwID can achieve greater independence through interdependence. In Lemon and Lemon's (2003) empirical study, they describe projects by community-based cooperatives in Toronto, such as Personal Futures Planning, microboards, co-housing and community work initiatives that demonstrate this approach and encourage those with disabilities to live interdependently and to participate as partners in their own businesses. In this way, they found that PwID are incorporated into the community, where 'their contributions, capacities, gifts, and fallibilities will allow a network of relationships involving work, recreation, friendship, support, and the political power of being a citizen' (Lemon and Lemon 2000: 57).

The approach is not so much to help them toward independence, but rather an interdependent life in the community where there are mutual benefits to those with and without disabilities. The following section maps out the potential geographies of interdependence identifying its possibilities and barriers.

An Interdependent Life: Possibilities and Barriers

With this interdependent paradigm advocated above, care has to be central to the social fabric of society and a key reference point in policy. Here the focus is on creating social relations with others, not dependence. Moreover, the focus is on creating a way in which we support what the person wants for a meaningful life; not normalization into predefined roles. The care of PwID should, in theory, inevitably lead to spaces where the person lives their life without the sense of being cared for. Social role protagonists therefore must be conscious of not simply promoting narrowly defined socially valued adult roles – as in getting a job, going to church, volunteering – as being the final goal. People have to be given real choice in what they want to do in life. However, with the presence of cognitive impairment, this process of supporting genuine choice and control is more complex.

In reality, according to Dittmeier (2009), choice is firstly *developmental*, in that people do not learn how to make good choices overnight. Secondly, choice is an *interdependent activity*, in that people typically rely on trusted allies to help them make good decisions. Thirdly, choice involves *complex options*, meaning most genuine choices are not a simple 'yes/no', 'either/or' or 'right/wrong' proposition; and finally choice implies *having options available*. In other words, PwID will not have genuine choices unless those in their caring nexus embrace service strategies that actually afford personalized options to be made, for example, where, who, how and when. Therefore a closer examination of the complex process of advocating and promoting a PwID's choices is needed, particularly in light of the recent drive towards person-centred planning.

In order to deal with the ever-increasing amount of wider significant relationships, person-centred planning (PCP) approaches have come to dominate the rhetoric associated with the design and delivery of residential, vocational, educational and recreational supports for adults with ID. It has been used as a way of enabling people to take a lead in planning all aspects of how the service they receive are delivered. Statutory services in England and many states in the US are required to introduce PCP as a means of increasing the extent to which supports are tailored to the needs and aspirations of adults with ID.

A number of different models have been developed in different places for the implementation of PCP. Each has a particular approach that is appropriate for certain individuals in certain situations. For example, in the United States, PATH (Planning Alternative Tomorrows with Hope) is used for the development of individual action plans, whereas Personal Futures Planning focuses less on services and tends towards building relationships with family, friends and the wider community (see Sanderson et al. 1997). The latter is often used for those who can gain a large amount of independence yet their families may often hinder the choices and control over their decisions. In order to make this move to encourage the growth of informal networks of support, Max (2007) makes the call for a shift away from a PCP facilitator to a Person-Centred Thinking Coach. The key difference is a change from 'process expert' to 'process sharer' with an emphasis on 'coaching' people's circles of support and people at every level of services in how to use them (Max 2007). Similarly, in Canada, an organization called PLAN Institute provides a facilitated social network development service, where the family hire a community connector to nurture the development of a personal network of committed people who join together in a relationship with the individual and with each other. The community connector approaches people in local mainstream settings to join the network based on the PwID's personal preferences such as the local horse riding club or nursery (Cammack and Etmanski 2002).

Another example of new forms of PCP is a 'social interpreter', which is currently being advocated in New Zealand to denote a shift from giving service to being of support in decision-making (Ferguson and O'Brien 2005). The concept of social interpreter recognizes that people with cognitive impairment often need support to understand an increasingly complex cognitive world, within which it is sometimes quicker for staff, friends and family to 'do for a person' rather than guide the person through the decision-making process. This concept is no different from the way in which people in general seek the support of friends and family to work through a problem. In essence, a social interpreter is an ally, and could be a friend, full or part-time support worker, or family member.

A holistic and ecological approach across the lifecourse is imperative in an analysis of these new forms of support, as allied care needs to be considered as integral to the development of a person's life, including the input of agencies such as housing, education and learning, urban and rural planning, and health (Philips 2007). By not assuming the family is the sole source in the care relationship, network methodology enables us to look at wider significant relationships. With group initiatives such as those mentioned above, PwID are gaining independence by relying on one another, on their families, on agency help, and interacting regularly with many people every day. Thus independence must be based on interdependence or it may lead to isolation.

To appreciate the geographical impacts of these new approaches on people's lives, McKie, Gregory and Bowlby's (2002) concept of 'caringscapes' is useful in conceptualizing these nested *inter*dependencies in a spatial way. A caringscape perspective would consider the complexity of spatial-temporal frameworks and reflect a range of activities, feelings and reflective positions in the routes people map and shape through caring and working. It requires the examination of the actualities and possibilities of the social patterning of time-space trajectories through a range of locales significant to caring. Thus an appreciation of the intermeshing of time-space uses, services, people and policies must be the basis for the conceptualization of the caringscape for a PwID. Geographers can try to map routes through the shifting and changing multi-dimensional terrain that comprises this new vision of caring possibilities and obligations (McKie, Gregory and Bowlby 2002). Within a caringscape, there is an acknowledgement of both segregated and integrated settings and a recognition that a mix of both can be effective because it recognizes the need of people with disabilities for contact with others who experience the same challenges they face, as well as for opportunities to participate in the mainstream of daily life (Lemon and Lemon 2003).

Empirical work has uncovered many place-specific caringscapes embedded in local geographies around the world. One example is Te Roopu Taurima O Manukau Trust, a Maori approach to support (see Tenari 2005). Its philosophy is based on developing a whānau, a Maori word for extended family. It employs elders to travel up and down the country to find the whānau of the people who had been referred to their service and re-establish broken ties with extended family. Unlike individualist interpretations of independence, they use the concept of rangatiratanga, which means a sense of autonomy within the context of an alliance with the members of the whānau. In other words, self-direction is a collective achievement, not an individual one.

While new approaches in PCP, such as the examples above, are offering real positive socio-spatial changes, in terms of developing social networks, contact with family and friends, and community-based activities (see Robertson et al. 2007); there are still many factors which prompt us to remember that we are often a long way from a caringscape perspective. This is especially the case when the choices are restricted by the parameters of social class and income (McKie, Gregory and Bowlby 2002) and the continued presumption of the seemingly naturalness of dependency of PwID. Indeed, when the contextualization of interaction is considered, the power relations of particular social practices within a locale and the ways in which they are achieved and maintained is important. The power differential is a crucial element in any re-analysis. A geography of ID must acknowledge this differential between providers and recipients of services. It should be challenged as it continues to enforce dependency and patronize disabled people. For many, at least some caring pathways are ill-defined, taken for granted or restricted by the availability of caring resources.

More generally, even though the policy focus of PCP is broadly accepted as the way forward for service provision, it has proved easier to talk about than to undertake. Recent evaluations (Dowling 2006, Ritchie 2002), which have critiqued PCP, have noted that there is a misguided reliance on PCP to do the system's work. They note that adherence to the underlying principles of PCP is not sufficient to develop interdependence alone; often the best planning in the world is unable to deflect the forces constraining the person's life. Furthermore, Robertson et al.'s (2007) analysis found that implementation of PCP is often described as partial or slow due to the slow pace of change in service culture and power relations, immutable funding structures, services' inflexible infrastructures, and high levels of staff turnover.

It is clear, also, that the introduction of PCP does not have an equal impact for all participants and in all regions. Place it would appear, still matters in terms of access and efficacy of PCP. Geographic factors are generally associated with the differential impact of the PCP process. For example, according to McIntosh and Sanderson (2005), one site which they identified as demonstrating good practice had several facilitators from a Citizen Advocacy service, who showed confidence and flair in helping people make real and positive changes. These underlying values appeared to be a strong factor in achieving change with people. This indicates a strong influence of factors relating to the contextual factors of PCP on both access to and the efficacy of PCP. Within this interplay, the impact of human agency can be seen to have a clear role in the creation of the specific geography of PCP provision.

Policy and research must recognize the complexity of a PwID's unique capacity to live independently, and draw upon this as a basis for developing and reviewing policies and services. The fact remains that PwID are one of the most marginalized groups in Western society, with epidemiological studies consistently showing a significant association between poverty and the prevalence of ID (Emerson 2007). In addition, PwID still have to live with the negative consequences of stigmatizing social practices and attitudes despite the grounding and extension of supports that enable them to live more independent lives.

In order to effect real change, PCP needs a fundamental change in the culture that permeates services. This includes an overhaul of services which have inherited resource systems that are managed and allocated on a whole-service basis without reference to the individual. Moreover, local authorities must be challenged to redevelop care assessments in a way that does not simply present a list of pre-defined services and benefits. The ethos must move away from providing a service to being of support. This includes a new breed of trained, confident staff, and multi-agency working which can engage with the person in working for what they want outside the system. At the same time, staff must encourage the growth of informal networks of support and develop allied links with mainstream services to be more accessible to PwID.

One way of shining light on the geographies of interdependencies is through appropriate empirical research. One of the major contributions geography can make is to examine how the complex issues of interdependence, control, choice and self-governance play out across space. For example, as Hall (2005) demonstrates, many people with ID are in a 'double-bind' of marginalization, experiencing exclusion from and abjection and discrimination within the very social spaces that are the key markers of social inclusion policy. It is therefore important to incorporate relational aspects of places through empirical studies in order to unpack whether the geographies of a person with ID are exclusionary or not. In perhaps many cases, the most suitable environment may be a combination of those invisible places of safety or networks of safe havens despite being 'out of fashion' in policy terms (Pinfold 2000, Hall 2005) and regaining use of mainstream places. Of key relevance is the degree of control, choice and self-governance that a person with ID has over their life. These are the indicators of achieving more independence. Some research is beginning to look beyond the narrow binaries of independence/ dependence and inclusion/segregation, employing appropriate research methods to reveal the experiences and hopes of PwID. At the same time, community action projects (see Honess 2007, for an example) must also play a leading role in kick-starting a series of conversations about how neighbourhoods can become safer places for people with ID, and to design conversations to enable others to listen to the views of PwID.

Conclusion

This chapter was concerned with the geographies of interdependency of people with intellectual disabilities. It has attempted to provide a more nuanced account of the fluidity and complex praxis of caring for this group. Stating that carers *cause* dependency negates the opportunity to develop a closer examination of care in which it is possible to see, even within the carer and cared for dyad, that parents grapple with and are confronted with conceptions of independent living. It is clear that care involves reciprocity and interdependency.

The chapter developed a social ecology approach, which forces us to focus on the wider set of network relationships and the role of interpersonal relations between all parties involved in the caring nexus, such as siblings, members of the public, agencies, and mainstream local service providers. Outside of the caregiving dyad, it is possible to see the importance of a collective set of nested interdependencies where PwID can gain independence by living interdependently within the community, relying on one another, on their families, on agency help, and interacting regularly with many people every day.

It is clear that PwID have a complex geography (in)between these nested interdependencies. A closer examination suggests that there is a place for both 'segregated' environments such as day centres that are welcoming and not stressful and 'integrated' spaces like mainstream public places in the lives of PwID. A complex geography is thus perhaps a somewhat necessary part of the lives of PwID. There is a concern about criticizing traditional day centres; while it is true that many day services need to improve, such critiques carry the risk of throwing the baby out with the bathwater. In other words it is perhaps a mistake to get rid of 'traditional' segregated places for the fear of them appearing out of fashion, without providing suitable replacements.

The emphasis on independence, choice and control cannot be quarrelled with in any discussion regarding ID; they are worthy abstractions. However, they are abstractions nevertheless, meaningless unless they are defined in context. Throughout the chapter, space and place were seen to play a key role in understanding the context shaping how these interactions play out. The geographical issues of housing, transport, mobility, and experiences of public and segregated space, all affect the geographies of PwID. At the heart of this, there is a recognition of 'impairment effects'; the intermeshing of the bodily (mental) restriction of activity with the effects of disablism in shaping the lived experience of the person with the disability (Thomas 2007). Future geographic research in ID has a key role to play here, showing how these concepts play out across lived space and how the everyday context of place shapes the experiences of PwID.

References

Birenbaum, A. 1971. The mentally retarded child in the home and the family life-cycle. *Journal of Health and Social Behaviour*, 12, 55-65.

Cammack, V. and Etmanski, A. 2002. A family PLAN for the future of people with disabilities. *Transition – The Vanier Institute of the Family*, 32(1), 10-15.

Crooks, V.A., Dorn, M. and Wilton, R. 2008. Emerging scholarship in the geographies of disability. *Health and Place*, 14(4), 883-888.

Department of Health. 2001. *Valuing People: A New Strategy for Learning Disability for the 21st Century*. London: Department of Health.

Dittmeier, H.L. 2009. *Supporting Genuine Choice and Control*. Brothers of Care Training Workshop Series, Temple Gate, Clare, Ireland, 28 January.

Emerson, E. 2007. Poverty and people with intellectual disabilities. *Mental Retardation and Developmental Disabilities Research Reviews*, 13(2), 107-113.

Ferguson, P. and O'Brien, P. 2005. From giving service to being of service, in *Allies in Emancipation: From Providing a Service to Being of Support*, edited by P. O'Brien and M. Sullivan. Victoria: Thomson-Dunmore Press, 3-18.

Forrester-Jones, R., Carpenter, J., Coolen-Schrijnerà, P. et al. 2006. The social networks of people with intellectual disability living in the community 12 years after resettlement from long-stay hospitals. *Journal of Applied Research in Intellectual Disabilities*, 19(4), 285-295.

Forrester-Jones, R., Jones, S., Heason, S. and Di'Terlizzi, M. 2004. Supported employment: a route to social networks. *Journal of Applied Research in Intellectual Disabilities*, 17(3), 199-208.

Gleeson, B. and Kearns, R. 2001. Remoralising landscapes of care. *Environment and Planning D: Society and Space*, 19(1), 61-80.

Gray, D.E. 2002. 'Everybody just freezes. Everybody is just embarrassed': felt and enacted stigma among parents of children with high functioning autism. *Sociology of Health and Illness*, 24(6), 734-749.

Hall, E. 2005. The entangled geographies of social exclusion/inclusion for people with learning disabilities. *Health and Place*, 11(1), 107-115.

Hall, E. and Kearns, R. 2001. Making space for the 'intellectual' in geographies of disability. *Health and Place*, 7(3), 237-246.

Hassall, R., Rose, J. and McDonald, J. 2005. Parenting stress in mothers of children with an intellectual disability: the effects of parental cognitions in relation to child characteristics and family support. *Journal of Intellectual Disability Research*, 49(6), 405-18.

Hatzidimitriadou, E. and Forrester-Jones, R. 2002. *The Needs of Older People with Learning Disabilities and Mental Health Difficulties in the Medway Area. Report for Medway Age Concern Groups*. Canterbury: Tizard Centre.

Honess, J. 2007. 'Keep safe' coffee house challenge. *Community Connecting*, 10 (September), 16-17.

Kittay, E. 1999. *Love, Labour: Essays on Women, Equality and Dependency*. New York: Routledge.

Laws, G. and Radford, J. 1998. Place, identity and disability: narratives of intellectually disabled people in Toronto, in *Putting Health into Place: Landscape, Identity and Well-Being*, edited by W. Gesler. Syracuse: Syracuse University Press, 77-101.

Lemon, C. and Lemon, J. 2003. Community-based cooperative ventures for adults with intellectual disabilities. *The Canadian Geographer*, 47(4), 414-428.

Max, N. 2007. From PCP facilitator to person centred thinking coach: from 'process expert' to 'process sharer'. *Community Connecting*, 10 (September), 10-12.

McIntosh, B. and Sanderson, H. 2005. Supporting the development of Person Centred Planning, in *The Impact of Person Centred Planning*, edited by J. Robertson et al. Lancaster: Institute for Health Research, Lancaster University, 13-23.

McKie, L., Gregory, S. and Bowlby, S. 2002. Shadow times: the temporal and spatial frameworks and experiences of caring and working. *Sociology*, 36(4), 897-924.

Milligan, C. 2001. *Geographies of Care: Space, Place and the Voluntary Sector*. Aldershot: Ashgate.

Morris, J. 1993. *Independent Lives*. Basingstoke: Macmillan.

Mulroy, S., Robertson, L., Aiberti, K., Leonard, H. and Bower, C. 2008. The impact of having a sibling with an intellectual disability: parental perspectives in two disorders. *Journal of Intellectual Disability Research*, 52(3), 216-229.

Parr, H. 1997. Mental health, public space, and the city: questions of individual and collective access. *Environment and Planning D: Society and Space*, 15(4), 435-454.

Phillips, J. 2007. *Care: Key Concepts*. Cambridge: Polity Press.

Pinfold, V. 2000. 'Building up safe havens...all around the world': users' experiences of living in the community with mental health problems. *Health and Place*, 6(3), 201-212.

Power, A. 2008. Caring for independent lives: geographies of caring for young adults with intellectual disabilities. *Social Science and Medicine*, 67(5), 834-843.

Power, A. 2005. *Landscapes of Care: A Geographical Study of Informal Care and Care Support in Ireland using International Comparisons*. Unpublished PhD Thesis, Ireland: National University of Ireland Maynooth.

Pruchno, R.A. and Meeks, S. 2004. Health-related stress, affect, and depressive symptoms experienced by caregiving mothers of adults with a developmental disability. *Psychology and Aging*, 10(1), 64-75.

Read, J. 2000. *Disability, the Family and Society: Listening to Mothers*. Buckingham: Open University Press.

Ritchie, P. 2002. Modernising Services: Person-Centred Planning. Section 6: *Working for Inclusion Series*. London: The Sainsbury Centre for Mental Illness.

Robertson, J., Emerson, E., Hatton, C., Elliott, J., McIntosh, B. and Swift, P. 2007. Person-centred planning: factors associated with successful outcomes for people with intellectual disabilities. *Journal of Intellectual Disability Research*, 51(3), 232-243.

Ryan, S. 2008. 'I used to worry about what other people thought but now I just think…well I don't care': shifting accounts of learning difficulties in public places. *Health and Place*, 23(3), 199-210.

Sanderson, H., Kennedy, J., Ritchie, P. and Goodwin, G. 1997. *People, Plans and Possibilities: Exploring Person Centred Planning*. Edinburgh: SHS Ltd.

Seltzer, M. and Heller, T. 1997. Families and caregiving across the life course: research advances on the influence of context. *Family Relations*, 46(4), 321-323.

Shearn, J. and Todd, S. 2000. Maternal employment and family responsibilities: the perspective of mothers of children with intellectual disabilities. *Journal of Applied Research in Intellectual Disabilities*, 13(3), 109-131.

Simpson, M.K. 2007. Community-based day services for adults with intellectual disabilities in the United Kingdom: a review and discussion. *Journal of Policy and Practice in Intellectual Disabilities*, 4(4), 235-240.

Swain, J., French, S. and Cameron, C. 2003. *Controversial Issues in a Disabling Society*. Buckingham: Open University Press.

Tenari, T. 2005. Te Roopu Taurima – a Maori approach to support, in *Allies in Emancipation: Shifting from Providing Service to Being of Support*, edited by P. O'Brien, and M. Sullivan. Victoria: Thomson-Dunmore Press, 129-134.

Thomas, C. 2007. *Sociologies of Disability and Illness: Contested Ideas in Disability Studies and Medical Sociology*. Basingstoke: Palgrave Macmillan.

Valentine, G. 1996. Angels and devils: moral landscapes of childhood. *Environment and Planning D: Society and Space*, 14(5), 581-599.

Watson, N., McKie, L., Hughes, B., Hopkins, D. and Gregory, S. 2004. (Inter)dependence, needs and care: the potential for disability and feminist theorists to develop an emancipatory model. *Sociology*, 38(2), 331-350.

Wiles, J. 2003. Daily geographies of caregivers: mobility, routine, scale. *Social Science and Medicine*, 57(7), 1307-1325.

Chapter 7

563 Miles: A Matter of Distance in Long-Distance Caring by Siblings of Siblings with Intellectual and Developmental Disabilities

Deborah Metzel

Introduction

My older sister, born in 1948 with Down's Syndrome, was never institutionalized and we were raised together at home. Since I began university in the late 1960s, we have lived apart; Charlotte[1] has always lived in Ohio. In 1987 my sister moved from our family home to a group home about 20 minutes away when our mother became increasingly ill. Our mother died in 1989, and when our father became much less able in his later years in the mid-1990s, I became more active in my sister's state-funded life. Now Charlotte and I are the only immediate remaining family members. She is her own guardian since the need has not arisen for one to be legally appointed.

In 2000, at Charlotte's annual Individual Service Planning (ISP) meeting, I asked our dad not to interfere or disagree with me about her desire to quit a boring, noisy, segregated, 'senior' day programme so that she could spend her time as she chose at the group home. Uncharacteristically, he did as I asked and remained quiet throughout the entire meeting. That was when I knew the responsibility for my sister had shifted from him to me and it was a sad, conscious or unconscious, concession on his part. Soon after, I became his long-distance caregiver as well.

I do not recall exactly when I began to fly more frequently to Ohio to take care of things for my dad and Char. After our dad died in 2005, my sister and I began to see each other about six times a year, with her flying to me more often than I to her. I purposely booked her on the same direct flight every time since the flight attendants got to know her as did the on-ground airline agents. A closet in my home contains a week's worth of clothes for her so that she need only bring a carry-on bag. One of my usual trips to Cleveland has been for Char's annual ISP meeting and while not mandatory for me to attend, I would never miss one since that is when requests are documented, previous plans changed and updated,

---

1 My sister gave me verbal permission to use her name in this work.

and service providers are held accountable. For my convenience, this meeting is scheduled to coincide with her birthday so that we can celebrate together.

Since Charlotte does not read or write, we do not send letters or notes to each other, although I mail the e-tickets for her trips to me. We used to phone each other about once a week. She has always been dependent on others to help her make phone calls and she usually called me when there was something in particular that she wanted to talk about, when she was bored, or to let me know that she received the e-tickets I had sent her.

As she approached her 60th birthday, Charlotte became less able and less adventurous. She started misspelling her name when signing cards or papers, asking the same questions in the same conversation, and getting lost in her daydreams, fantasies, and tasks more often. Since mid-2008, I began to fly more frequently to Ohio to see her or to fetch her for a trip to Massachusetts because she prefers me to travel with her. Regardless if her decline is due to ageing or incipient Alzheimer's disease, it distresses me because the end result will leave her at a place much more difficult for me to reach than her current physical location 563 miles away.

My own situation as a long-distance adult carer sibling for my sister with an intellectual/developmental[2] disability (IDD) has entirely motivated this research. My self-proclaimed expertise as a long-distance carer is meant to explicitly establish my position in this research while my life as a geographer enables me to question and understand the spatiality of our relationship.

This chapter aims to explore the spatiality of being an adult sibling carer of adult sisters and brothers with IDD. This first pass through sibling long-distance caring has several intentions to redress our invisibility in informal caring. One is to begin to remedy the relative lack of 'subdisciplinary writing about…the practice of caring' (Parr 2003: 216), especially as practiced by long-distance carers. Through the tangible and intangible experiences of us carers, I mean to provide 'important empirical details about caring roles, practices, knowledges, and relationships as they are negotiated and played out within and through space' (Parr and Philo 2003: 472), particularly as they have occurred over long distances. Lastly, this chapter intends to reveal more about who we are because of our distance from our siblings. Ultimately I want to begin to discover how our long-distance caring relationships with our sisters and brothers transcend space and how this durable spatial relationship affects us.

---

2   In the mid-2000s, the term 'mental retardation and developmental disabilities' began to be replaced by the term 'intellectual disability' in the United States. According to the American Association on Intellectual and Developmental Disabilities (AAIDD) (2009), developmental disabilities refers to physical as well as intellectual disabilities and may also include Autism Spectrum disorders. For an excellent overview of the change in terms, please read *FAQ on Intellectual Disability* at http://aaidd.org/content_104.cfm.

Who and Where are the Carers?

Caring is a 'loaded' concept, 'a complicated social act' (Parr and Philo 2003: 471) involving at least two people: the care-giver and the recipient of such care who has been socially constructed, by others and/or by themselves, to be care-recipients. Much of the research on informal, unpaid family caregiving has explicitly and implicitly placed caregiving in close proximity, with the caregiver and care recipient often sharing a home (cf. Joseph and Hallman 1996, 1998). Typically, informal caring family relationships are those of parents of children, children of ageing parents, and couples, with the latter implied as being spouses or significant others. Friends and neighbours are other carers whose proximity frequently brings them into the care recipients' homes. From my long-distance location, I am compelled to ask: 'Where are we siblings in the study of informal caregiving?'

In the wake of the first wave of research on family caring for ageing and declining parents, research on adult sibling care of ageing and declining siblings began to appear. It has revealed the demographics of the caregivers, and the impacts and costs, both psychological and financial of long-distance caregiving (Ericksen and Gerstel 2002, Koerin and Harrigan 2002, Seltzer et al. 1997), and the relationship during the different life-stages of adult siblings and the degree and type of caregiving for siblings (Eriksen and Gerstel 2002, Seltzer et al. 2005). As in other informal caregiving relationships, women compose the majority of carers. Sociologists Eriksen and Gerstel conclude that 'alongside mothers and daughters... sister work helps to create a familial sense of place, meaning, and ritual, often across considerable distance' (2002: 852), which is some acknowledgement of the strength of sibling caring relationships over space.

Early on Seltzer et al. placed us in:

> a pivotal position within the family, despite the fact that they generally do not live in the parental home, typically have their own family and employment responsibilities, and are infrequently involved in daily care-giving for their dependent brother or sister. Rather, their role appears to revolve more predominantly around the provision of affective support to their adult brother or sister (1991: 316).

Despite the importance that has been assigned to us by researchers in the field of IDD, the flurry of research in the 1990s has not grown much in recent years. In 2005 a focused section on siblings in the journal *Mental Retardation* included the reminder that the 'field of sibling studies was still underdeveloped...' (Hodapp, Glidden and Kaiser 2005: 334).

Unlike the slow or sudden changes that prompt long-distance caring for ageing parents or adult siblings, most of us grew up with our disabled siblings 'whose disability was a prominent feature of family life' (Greenberg et al. 1999: 1214). In the US, we are now the adult siblings of more than 710,000 adults with intellectual disabilities who are living at home with parents over the age of 60 (Swenson 2005:

368 citing Rizzolo et al. 2004); the number of us whose siblings live outside of the family home is unknown.

Carer siblings have been described as generally supportive, concerned, and involved (Seltzer et al. 2005: 355). Stoneman (2005) inventoried other qualities we share, some laudable, some not. We have both positive and negative feelings about our disabled siblings that 'fall within the realm of healthy emotional experiences' (Dykens 2005: 362). Similar to a group of adult carer children of ageing parents described by Roff et al. (2007: 329), carer sisters and brothers are often educated beyond high school (Krauss et al. 1996, Greenberg et al. 1999; Rimmerman and Raif 2001) with many of us working in human services or academia (Gorelick 1996, Roff et al. 2007: 329). In contrast to relationships between non-disabled siblings that may include a strong degree of friendship (Eriksen and Gerstel 2002: 839), we tend to act as surrogate parents to our siblings (e.g., Bigby 1997, Seltzer et al. 1997), which sets up the immutable emotional, but not spatial, boundaries of our relationships. We and/or our parents have expectations of our future responsibility for our siblings (e.g., Greenberg et al. 1999, Krauss et al. 1996, Rimmerman and Raif 2001). Swenson (2005: 365) notes formal human service systems also 'rely on an assumption that…siblings will step in later to do what parents have always done', which I argue, once again assumes proximity.

However, a number of us do not live in close proximity to our siblings, yet there is little information available on such long-distance carers. In a study on plans for future involvement with IDD siblings, Krauss et al. found that of the non-disabled siblings who did not live in the family home, 58 per cent 'lived within a 30 minute drive from their parents' home' (1996: 88). No other measure beyond 30 minutes is mentioned, suggesting the importance of proximity and leaving the distance beyond this measure tantalizingly mysterious. In another 1996 publication, the statement that 'perhaps 30 per cent (of non-disabled siblings)…' made a purposeful choice to move 'as far away as possible from their (IDD) sibling' (Goreflick 1996: 7) is not illuminated by any other details.

While this section brings us siblings into the frame of informal caring, it does little to enlighten us about the role that space has in our particularly physically distant relationships with our siblings with IDD. For that I call upon geography.

The Power of Geography in Caring and Caregiving

Social geographical concepts are best suited to direct an investigation of the spatiality of our long-distance caring relationships with our siblings. In particular, the conceptual framework introduced by Dear and Wolch underpins this research. They recognize three aspects to the socio-spatial dialectic:

  (1) in which social relations are constituted through space;
  (2) in which social relations are constrained by space; and
  (3) in which social relations are mediated by space (1989: 9).

Distance is the fundamentally physically defining element of our current sibling relationships, creating the social space in which we siblings care for our sisters and brothers. The current body of research on long-distance caring provides a small platform from which to investigate long-distance spatial relationships with our IDD siblings, with its explicit and implicit focus on proximity in informal caring. Until recently, the socio-spatial dimensions of proximate informal caregiving relationships were obscured. Milligan uncovered the constitution of trespass through the intrusion of formal care that weakens the informal carer's sense of control of the home space and changes the space from 'private/domestic to public/ domestic...' (Milligan 2000: 55). Others have written about the socio-spatial constraints on the daily lives of the care-givers (Hallman and Joseph 1997, Joseph and Hallman 1996, 1998, Milligan 2000, Wiles 2003).

Geographers have failed to expand upon the role of distance in familial caregiving since its introduction by Joseph and Hallman in 1996 and 1998. Quantification does underscore the deep commitment to provide care over long distances in time and miles in caring for elderly parents[3] but this is inadequate in understanding the spatiality of these caring relationships. Joseph and Hallman found that adult children go to '"extreme" distances, perhaps regularly travelling [sic] five to six hours each way to provide care' (1996: 638), supporting the observation by Eriksen and Gerstel of the 'considerable distance' (2002: 852) in caregiving. Schoonover et al. provide a useful notion of long-distance that was 'determined by other research to be a threshold point at which visiting and face-to-face interaction...decreases significantly' (1988: 475).

In the field of intellectual/developmental disabilities, one article touches briefly on geography. The article, 'Who would I be without Danny?', recounts how the location of her brother Danny becomes a distinct, recognized, and integral part of a woman's identity (Flaton 2006). The participant acknowledges (2006: 142) that Danny has 'shaped her life with his needs...from her choice on *where to study* and what to study to *where to live*' (my italics). Danny and his sister live in the same state at an inconvenient distance from each other and they depend on their older brother to fetch Danny and bring him to their sister's home for weekend visits several times a month.

We are told that we are important to our siblings with IDD and we have grown up with part of our identities shaped by our siblings. As we, and our sisters, brothers

---

3 The Alzheimer's Association LA and Riverside (2002), Lawton, Silverstein and Bengston (1994), and MetLife Mature Market Institute (2004) provide some data on distances between adult children carers and their parents. Regarding operational measures, Schoonover et al. used a 50-mile measure for 'geographically distant' (1988: 475) while a one-hour measure was used by the National Council on Aging (Koerin and Harrigan (2002: 66 citing Wagner 1997a). In Eriksen and Gerstel's study on adult sibling caring of non-disabled siblings, they found 'whereas 37 per cent of respondents' siblings lived in the same city or within 50 miles, the same percentage lived well outside the sample region, underscoring the wide range of sibling ties tapped by this particular study' (2002: 842).

and parents, age, we continue to be understudied at a time when an increasing number of us are becoming caregivers for our siblings. We are overlooked on several important fronts: as siblings and as informal carers, and so we have joined the ranks of the many other invisible and unheard informal family caregivers (Grant 2007, Parr 2003). Distance has also obscured us.

Methodology

I am both the researcher and the researched. I am intrinsically situated in this research because Charlotte and I live so far apart which constitutes my long-distance caring relationship. I claim responsibility for her regardless of my geographical and legal residence. I have lived this relationship for several decades, leading me to use a phenomenological approach to gain entry to other siblings' spaces and identities. The very characteristic of subjectivity in phenomenological research is what yields the important socio-spatial meanings of our long-distance caring relationships. Twenty years ago Ferguson and Asch argued for scholarly acceptance of work informed by what they termed 'personal narrative', meaning a 'published account of life with a disability written by a disabled individual or the parent of such an individual' (1989: 109) because of the authenticity of the experiences unshaped by the researchers' perspectives. Now autobiographical work is more accepted. Moss discusses the use of autobiographical writing as not only a means for self-reflection, but as the means to reach a 'fuller understanding about a specific experience' (1999: 163) which permits subjectivity and self-reflection in comprehending the long-distance sibling caring relationship.

My position in this work is an important advantage, not only as a motivation and lens for this work, but for its importance to the sisters and brothers I interviewed. With this group of participants, I felt assured that they understood that I was first a long-distance sibling, and secondly a researcher, and many of them commented on my 'getting it' when they spoke about their long-distance caring. I have striven to avoid a 'routine, and perhaps meaningless' claim of membership to a group under study (Butler 2001: 271) by writing myself into much of this chapter with my empirical and emotional experiences. Personally, I would be less inclined to accept research on sibling caring relationships by researchers without this experience precisely because the lived experience is the necessary background.

As a participant in my own research, I interpreted others' experiences along with my own. My perceptions, experiences, knowledge, and understandings as a long-distance carer sibling compose only one set of data among the many that inform the analysis. Other siblings and readers may interpret my findings somewhat differently, but few would argue that space and distance are not profoundly involved.

Operationally I had to choose a threshold distance as a criterion for this work and I used two hours by any travel mode as an inconvenient distance. So that the participants would describe their experiences of caring and care-giving without preconceived notions, I intentionally did not define the terms 'caring' nor 'caregiving', which both the participants and I used interchangeably.

I recruited through service-provider organizations, listserves, and by posting notices on my campus. I also used the snowball strategy. Twenty-three participants gave written informed consent and pseudonyms are used for all these participants and their siblings. Specific locations have been changed or omitted to preserve privacy. Interviews were conducted in-person or by telephone, and recorded. Background literature and my own experiences guided the questions in the semi-structured interviews. Given the manageable number of interviews, and my preference for reading and re-reading the interviews, I coded them myself. As codes developed into broader themes, I began to write and analyse, and I continued to interview and code. Through contrast and comparison and partitioning of variables (Huberman and Miles 1998: 187), I found a set of common themes.

As with any research, there are a number of limitations to this work. Participants self-selected. It is unknown how these findings are generalizable to other informal or sibling caring situations, but this work generates knowledge that can be, and should be, further examined, challenged, and expanded upon.

Findings

Participants' ages were from the mid-20s to late 70s, with the majority in their 30s, 40s and 50s. Participants and their siblings had legal residence in North America. Most of our siblings were not living in the family home with our parents. The total number of children in the families ranged from two to five. Many of our parents had died or were elderly and/or ill. The majority of the participants were women, although two men participated. The majority of participants were in human services or academia.

Selected findings are organized along the three aspects of the socio-spatial framework: constitution, constraints and mediation. Most of them fall neatly into one of the three aspects, but due to the innate complexities, spatial and otherwise, of these sibling relationships, readers may see alternate classifications. Regardless, what is obvious is that long-distance caring relationships can be analysed by this framework illustrating the power of geography.

Constitution and Reconstitution of the Relationship

The most significant location in this relationship is that of the disabled sibling. No matter the living arrangement – a family home with one or both parents, and/or other siblings, nursing homes, small institutions, group homes, residential treatment programmes, and apartments with various degrees of support by provider agencies – the sibling is the locus of a sphere of influence that establishes who has the responsibility for taking care of the disabled family member. While the majority of the participants grew up, as I did, with our siblings in the family home, some did not. Nonetheless, at some point in time each accepted responsibility for their sister or brother and continue to care for them.

Statutory Control

Statutory controls are an important factor in informal caregiving (Milligan 2001) and can directly control the degree of legal involvement by us carer siblings. In the United States, two statutes can interfere with our involvement. First, guardianship of any person 18 years old and older must be court-appointed, and states and provinces vary in their residential requirements for the guardian. Ohio law excludes me from being appointed as Charlotte's legal guardian[4] because my legal residence is in Massachusetts. As Charlotte's dementia continues, I anticipate having to reckon with this important spatial dilemma. Another federal law, Health Insurance Portability and Accountability Act (HIPAA) of 1996 (P.L.104-191), maintains the privacy of and accessibility to an individual's health care information, thus preventing carer siblings from obtaining information about the disabled sibling unless the carer sibling is the legally appointed guardian.

> Carol: I was accepted as acting on (my mother's) behalf whenever I was around and whenever things needed talking over…Once my mother died…we have had trouble because a great deal is made of the fact that I do not live (there) so I do not have real entitlement to be involved in the way that my family wants me to be.

> Trina: I notice…like in my involvement in the last few years, interacting at a distance with the mental health system…my relationship as a sibling doesn't carry as much weight as my parents' relationship to him as parents. And at a distance I find that even harder. I would call the psychiatrist and instead of calling me back, he'd call my mom.

But guardianship is not without its price. Pam is not comfortable about the trade-off of access to her brother's information and his personal rights:

> I don't really like being his guardian. I think it violates his personhood, but I asked to be his guardian in order for people to talk to me, because if I wasn't his guardian and I was just his sister, they wouldn't have to talk to me.

Family Control

Especially when the disabled sibling lives with one or both parents, distance has been used to assert parental control. Though parents may not have legal guardianship of their adult child, their proximity reinforces the primacy of their

---

4   However, legally I am her health care proxy.

familial position. Some carer siblings feel dismissed by their parents when their assistance has been rejected on the basis of distance.

> Carmen: I think that was around the time I was trying to get him plugged back into the system and I didn't tell my parents because my mother took everything then as an insult...she would say things to me like '(You) don't understand, (you're) not there, (you) don't see it everyday, (you) don't live with it everyday, day in and day out, (you) don't understand how difficult he is and how little we can do...'

Like so much in geography, the scale of observation shifts how the disabled family member is viewed. Often, the proximate, frequent, and fine-grained parental view is that of a more dependent adult child in contrast to a sibling's distant and broad view of a more capable disabled sibling.

> Nina: I think he's capable of more independence than my parents do and I have more faith in him and I think he should be able to take risks.

> Carmen: I would say things like (to my parents)... 'I think you're babying him...'.

The disparate spatial perspectives often create conflict between the parents and the long-distance sibling and further contribute to the other frustrations and complexities in our long-distance caring relationships.

Reconstituting the Relationship

Home landscapes, loaded with individual and family histories, cue us to intervene or advocate on behalf of our disabled siblings with our parents or to act as surrogate parents. We assume or try to resume our traditional role of carers when we are on home ground. When we lose the physical space of the family home, we have to find new spaces in which to maintain our relationships.

Soon after our dad died, I sold our family home, and though neither my sister nor I had lived in it for decades, we lost our shared sense of place in this truly familiar place. Over time we claimed several new comfortable, familiar spaces for ourselves that have proved durable. In Ohio, our territory is fixed by a number of points where we spend time together. These mainly include the assisted living facility where our aunt lives, two favourite restaurants, my hotel room, and the cemetery where many of our family are buried.

Not surprisingly, neither Char nor I consider the group home where she has lived for over 20 years, a home space for us together. The extra-familial home of the disabled siblings cannot be genuine home base for the pair of siblings when only one sibling lives there. Other long-distance siblings spoke about the lack of a home space. When Ricki and her husband visit her brother Jim, the three of them spend time together in the hotel room. Jim watches sports on the TV while Ricki reads and her husband uses his computer. Jim and Ricki are content to be in each other's company, although she says it's not the same as it would be in the family home. Jennifer found that 'not having a place where John and I can be together' made spending time with him more stressful. One of the carer siblings spoke about staying at her sister's group home because it actually had a guest room, but during those visits, Lara is a guest in what is clearly Shelly's home.

Periodic and ordinary errands are pilgrimages that comfort us and make otherwise inconsequential places important places. Even though care workers at Phoebe's home take her shopping weekly, Ted's visits always include an extended shopping trip to the local drugstore where she buys the sundries *du jour*. No matter if we are in Ohio or Massachusetts, Charlotte and I always go to a nail salon for pedicures and to an office supply store so that she can stock up on masking tape and cellophane tape for her hobbies. I believe that the ritual of these small excursions compensate somewhat for the lack of more historically shared spaces.

The Constraints of Distance

We carer siblings are greatly thwarted by distance since it limits access to information about our siblings. Any problem or concern about our siblings heightens our worries about their situation, especially since we are not there to fix it. Particularly when it comes to health problems and personal care, we carer siblings get upset and angry when care workers fail to do their jobs.

> Bob: She became very lethargic and wasn't eating…It turns out her hands were shaking and she couldn't get anything to her mouth…It shouldn't be my having to tell people in the profession what's staring them in the face.

> Marlo: Yeah, but sometimes I get so angry, like why do I have to tell you six times that she needs to have her hair done? You know, there were weeks I was writing notes, I was calling, I was leaving messages.

Distance breeds mistrust. When staff members do not take care of obvious and simple situations, our imaginations run riot with other possibilities of neglect. We are apprehensive about staff when we have little contact with them and when staff turnover is so high. We are rarely 'there' enough to begin to get to know the different carers on a more personal level, although that is an advisable strategy.

> Ricki: I know there was somebody my mother didn't trust as far as she could throw this person, she was so happy this person left. After my mother's death, this person came back and I don't know what to think, I mean, I know my mother's report, I have a little different sense of that, I don't know who to trust. And I worry about abuse, I mean I don't know – how do you know? How do you know? I don't like some of the things I see, but I don't know – how do you know?

We do not know who to trust or how much to trust anyone who is working with our siblings. Distance complicates our communication with care staff and/or agency staff.

> Carol: Any person I was able to get in touch with would be at some…desk job, who I'd never met, had nothing to contribute to the situation, and the agency running (the) group home…had policies that prohibited care (workers) from calling long-distance. So I made numerous efforts to set up a system by which they could call me with a phone card, a number of different ways, and all of them met with resistance, then finally an administrator said that they were going to allow calls to certain numbers and my number was on that list but I've never received a phone call.

> Pam: (T)he second (difficulty) is dealing with all the systems that are supposedly in place to care for him…And the system is just so screwy that they don't talk to each other, they don't tell the truth, and half the time no one knows what's going on, so it's hard to keep information straight. You know, they write down all these progress notes, whether it's happening or not, when you're long-distance…

Distance further reduces our already limited power to take care of our siblings. Without being close by to see for ourselves, our anxieties are heightened and we become more critical of carers, and ultimately ourselves for not being there.

To be fair, there were a few siblings who were very satisfied with the care of their siblings. When Maeve's sister began treatment for breast cancer, Maeve was completely reassured that staff members at the home were taking good care of Shelly:

> They're taking her to radiation treatment everyday and they had a plan, you know. They did this great research and found out the American Cancer Society has a free shuttle bus…to take you to your appointment and bring you back again – isn't that great? So they found this out and they said okay, we can take her to a radiation treatment until she feels comfortable and then we'll just let the Cancer Society do it. But the house manager called me a couple weeks ago and she said, 'I know we had this plan but I can't do it, I can't let her go all by herself, you know…I can't just let her sit there by herself, I know she'd be fine but I wouldn't be'.

The house manager's discomfort at letting Shelly go by herself mirrored Maeve's concern and desire to be with her sister and so was appreciated and reassuring. More often, we are continually frustrated by distance in our care of our siblings. To compensate for not often being there ourselves, we mediate and manage distance as best as we can.

Mediating Distance

We long-distance sibling carers do not mediate the obstacle of space by any unusual means. We visit, we call, and we send messages and packages through the post. Email and other advanced technology are not commonly used since most of the disabled siblings in this research do not use them. Both the carer siblings and the disabled siblings prefer visits. Visits between us siblings fulfill different sibling agendas, but ultimately we each get what we want or need from seeing each other. Although I can detect when Charlotte is unhappy over the phone, I get much more information about her distress in person. In that moment of shared space, I automatically begin to figure out what to do next and Charlotte can see that I am trying to take care of her. There is simply no substitute for seeing each other face-to-face.

> Ted: Phoebe likes getting mail but she prefers in-person because basically, it's all in person.

> Nina: He's really prioritised face-to-face contact and visits. I think he fears that if he's talking on the phone to someone that they won't come to see him.

> Pam: If I was there, I could just go meet him, walk over. I could go to his house and see if...he needs new shoes and stuff like that.

We often travel to our sisters and brothers to manage or assist with medical care. Nina flew 13 hours to be with her brother during his hospitalization and felt that she was there for him in a way that their parents could not be. Glen was in the midst of managing his sister's surgery when he participated in this research:

> As the brother/guardian with remote oversight, I am leaving Tuesday night for 5 days in Minnesota as I authorised a full knee replacement surgery for my sister...After the 4 days in the hospital she is on to a nursing home rehab unit. Remote management of this level of stuff is difficult.

After a state-reimbursed dentist in Ohio recommended that Charlotte have all her teeth removed, I took over her dental care. For the past four years, she has had her annual dental exam and cleaning with my dentist and she returns to Massachusetts for any follow-up treatments.

For some siblings, face-to-face interaction is crucial when communication is limited. Carol's brother Mike communicates through sounds and gestures that require them to both be in the same space:

> He will point, he'll pull me by the arm or the hand...and he will lean his head in a certain direction and make some syllables...but without context, it's not dialogue at all.

As in other long-distance relationships, phone conversations mediate distance. Distance is not a factor in the frequency and number of calls, instead frequency and number of calls depend on both siblings, and especially if the disabled siblings needs help in calling their siblings as Charlotte does.

> Marlo: I talk to her on the phone, some weeks it might be everyday, and some weeks it might be, if I can go five days, depending on what my life is like. It's not like I feel compulsively like I have to call her every night.

Carer siblings described phone calls from their siblings as requests for emotional and social counsel and support, less so for money or specific items, thus confirming our supportive, affective role.

> Marsha: So, it's predominantly emotional support. She calls probably 3-4 times a week, to talk through things that have happened, and to sort of get a sense of how she might deal with people or social situations that she isn't sure how to handle. She has…some practical assistance nearby from folks like a job coach, but you know, if it's evenings or weekends or other times that she wants advice, she'll give me a call…So in some ways, it's a lot of helping her to strategise and helping her to interact with people so that she can maintain the supports that she has in place, and maintain her job, and her housing.

Ben phoned his sister to complain that their dad would only permit him one cup of coffee, not the five he wanted. Another time, he called when he was upset that their mother came into his room to take his dirty clothes to be washed. To Ben, his sister Nina is his advocate and likely to take his side in family conflicts. She soothes Ben and talks about these problems with him, and later speaks to their parents.

Our first response to our siblings is to help. Since we are so often called upon to help, phone calls from our siblings sometimes can confuse us even when our siblings may be relating information and not requiring our help.

> Darla: I think because Cheryl is not very good at communicating. While she does communicate things to me when she has a desire, sometimes she is not very good about getting her facts straight, and so whenever we have a situation…or something comes up, I always have to check with mom, find out what the real deal is, because I'm removed, sometimes I feel as though I will jump to conclusions based on something that Cheryl has told me. I can give you can example. Cheryl was at work at…a grocery store, and she said the power is out, so we're all in the break room but work is closed. I said, Oh, well, is mom going to pick you up? She said, I think so. The truth was she knew so. She knew mom was going to pick her up. Well, I said, Cheryl, get off the phone with me and call mom and make sure she knows what's going on…So, she called me back a few minutes later, and said, mom's coming in 20 minutes. Well, I came to find out later that she had already just spoken to mom, so basically she, because I prompted her to do that, she got confused.

The importance of phone calls becomes more evident when they lose their effectiveness. When our siblings become less able to communicate by phone, our level of concern and anxiety increases since one of our best ways to mediate distance is no longer possible. When Bob's sister's hands began to shake, she began to be unable to use the TTY at her nursing home, which further limited her communication with him. Ann also has begun to lose touch with her brother:

> This is probably one of the most frustrating things that I have ever experienced in my life because…he is now to the point where they think, probably Alzheimer's has started…And he has hearing difficulties, so talking on the phone doesn't work.

Writing and verbal communication are not easy for her brother, so Trina appreciates his 'very quirky way of communicating with me', evident in an 'entire shoe-box full of postcards, cards and collages' from him, though their receipt depends on his being able to mail these items to her. When Nina graduated with her doctorate degree, Ben, an artist, sent her a large painting, a very valued gift.

In summary, we care as best as we can over distance. We strive to make our presence, involvement, love, and caring known to our siblings and their carers.

Guilt and Caregiving: Does Distance Matter?

Despite our clear realization of the constraints of distance, our deep sense of responsibility often makes us feel guilty for not being there to do whatever needs to be done, to know whatever is happening, to take care of our siblings in person, to have them see that we are there taking care of them, and to just be with them.

> Trina: I'm always trying to explain it to people, that even (the) times when I'm not directly giving care, the caring part has me just as tied up. I feel just as shackled, sometimes, to the situation.

> Ricki: I feel guilty. I don't see him, but you know, four times a year, and that's usually 3-4 days – 3 days usually and it feels like not enough, it seems so unfair to him.

> Meg: I remember when I first went to college I felt really guilty when I first went (back home). When I was living at home, I was home a lot, my parents worked, so I would always be home after school to hang out with Theresa, and I remember when I first went to school, I got really upset at one point because I felt guilty that I wasn't there to hang out with her. I remember I called my mom once and I was like, we need to get Theresa more involved with things in our community and stuff, you know. I don't want her sitting at home without anything to do since I'm not there. So yeah, there was definitely a sense of guilt in the beginning, and I still feel that way sometimes.

To compensate for this, we are fiercely vigilant. Both the press of responsibility and the constraints of distance are evident in Bob' comment, 'If anything at all occurs, they are to call me 24 hours a day, so that takes care of that. I don't know what more I can really do, other than what I've done'. Bob verbally acknowledged the limits of what he can do, but his interview was filled with descriptions of his constant vigilance. His own declining health motivated him to legally appoint his daughter as co-guardian with him. She now makes the yearly long-distance trip to see her aunt and reports back to her father.

Stress and strain has been attributed to distance in long-distance caring relationships (e.g., Joseph and Hallman 1996, Koerin and Harrigan 2002, Schoonover et al. 1988). Disturbing feelings of guilt, frustration, and anxiety are omnipresent in informal caregiving of family members and in particular, distance has emerged as a cause of guilt and frustration in long-distance caregiving. In listing the 'additional strains' in long-distance caring, Koerin and Harrigan include 'the emotional burdens of anxiety and guilt that caregivers may experience because they are not as available as they feel they should be' (2002: 67). Roff et al. also attributed distance as a contributing factor in frustration and guilt in long-distance caregiving of parents (2007: 329, 330). The role of distance in constituting or contributing to guilt in long-distance caring of siblings with IDD is appreciably supported by the siblings in this research.

Is There Liberty in Distance?

As often and as deeply as we feel guilty, frustrated, and anxious because of distance from our sisters and brothers, we sometimes find relief in distance, and then guilt, of course, at our relief.

> Trina: Yeah, and again, it falls in that paradox, I think from me being a sibling, is really a living paradox, I mean at the same time, I feel guilty, want to be (there). If there was a crisis, I was relieved that I wasn't in the fire. When I was out of the country…that was a whole other level of distance, very, very interesting…I never experienced that level of 'I can't really do anything'. I mean, 'I'm not jumping on a plane right now. I really physically can't do anything right now', so I was left with my emotions in a different way.

> Ann: 'Then there's the whole psychological piece of being removed, and not having the constant reminder of anxiety, and responsibility and guilt in my life. And that probably makes no sense to anybody else'. She later said, '…(It's) not seeing the visible…I hate to say this, but the visible sign of what's for me, always triggers acute anxiety and guilt, right there in front of you, you know? Even in the community, if it was an hour away, I would still have that'.

> Lara: I don't feel like I have to be responsible for everything, I can say, 'I can only do so much here (where I live)'. I miss her because she brought a lot to my life but at the same time it was kind of nice not to have to deal with her all the time.

The same distance that provides temporary respite from responsibility also permits us to have identities independent of our disabled siblings. Now retired from a career in special education that was influenced by her brother, Ann says that if John lived nearby, 'I would be right back into my special ed mode again'. Others spoke about having their own space and identities away from their siblings.

> Nan: I sort of like having my own space…because here – especially since we grew up together, my brother went to the same high school I did and everybody knew him before they knew me, and sort of my community here, everybody knows me because they know him – so when I'm away (at medical school), I sort of get to know people, just, they don't know him, they just know me, and so that's kind of nice.

> Nina: If there's anything positive about not being in the same city with him, it's that I don't have as much conflict in deciding to have my own life.

> Trina: I don't know if I would have as much of my other self. The distance really provided me with that sense of other self and therefore I was more present with my other self in (my) marriage.

We so often think of ourselves in relation to our sisters and brothers that we can forget that we have other identities that are not linked to our siblings. Distance helps us to recognize that we are not only our siblings' carers.

Transcending Distance in Long-Distance Caring

At a distance, the most pronounced element in our self-identities is the responsibility we feel for our brothers and sisters. For many of us this has always been a fixed part of our identity, and our current long-distance relationships are just the latest version, supporting Heller, Caldwell and Factor's summary that '(a)dult siblings tend to maintain high levels of involvement with their sibling with disabilities across the life course' (2007: 138 citing Zeitlin 1986 and Seltzer et al. 1991). My research now indicates a high level of involvement across the life space.

A minority of sisters and brothers in this research did not grow up with their disabled siblings; nonetheless they took on long-distance caregiving of a barely known sister or brother at some point in their lives, forging some part of their identities with the assumption of this responsibility. Statutory regulations and parental control may block us, but we do not retreat from our commitments.

We are devoted, loving, vigilant, and often angry and guilt-stricken carers who have not only accepted responsibility to care for our siblings, but who have also accepted the difficulties of distance, as well as the freedom to be our other selves.

Yet in thinking about the spatiality of our long-distance caring relationships, I see that our responsibility, our commitments, transcend any physical distance. In thinking about the placelessness of our relationships, I wonder how I would comprehend the spatiality of my commitment to Charlotte if I lived close by. Would I feel more responsible or less responsible since we would have the luxury, and problems, of proximity? I cannot begin to know. When I consider the number of years we have spent apart, what strikes me as a/spatially significant is how the caregiving space operationally remains the same. This space is not bound by absolute measures of distance; the space is delimited by my sister and me, and what is important to both of us is what goes on in that space and despite that space.

Conclusion

This chapter examines the nature and affects of a being a long-distance sibling carer relationship using the socio-spatial conceptual framework introduced by Dear and Wolch (1989). Informed by interviews of other long-distance carer siblings and my own life as an adult long-distance carer, this chapter adds to the sub-disciplinary writing about caring and provides glimpses of authentic practices, strategies, and knowledge about how we care 'within and through space' (Parr and Philo 2003: 472). For the most part, our sisters and brothers are a distinct cohort as the first generation of people with IDD who were not routinely institutionalized. Our family histories and relationships with our disabled siblings in turn make us siblings a distinct cohort as well. In the realm of caregiving, we have been inconspicuous as siblings as well as informal caregivers, though we are growing increasingly visible and vocal. Distance also has diminished us. But we siblings are now claiming a place in geography by our increasing presence and actions as carers of our siblings.

The most pronounced element in our self-identities is our accepted responsibility for our siblings. No matter when we became caregivers, this is now a fixed part of our identity. Our current long-distance relationships are merely the latest adaptations of enduring relationships that wax and wane over time and space. Our physical locations define an operational caregiving space and for us long-distance carers it obviously creates problems, but no matter, we will take care of our sisters and brothers regardless of distance. As long as I have thought about my long-distance caring relationship with Charlotte, Goodchild's (2000: 84) definition of topology has stayed in my mind: 'Topology: In mathematics, a property that is invariant under spatial distortion'. This surely applies to our caring. In this meaningful way distance does not matter.

Figure 7.1   Charlotte and Deborah in 1957 in the photo booth at an amusement park in Cleveland, Ohio

Acknowledgements

Infinite gratitude and thanks is due all the carer siblings who participated in my research and, of course, to Charlotte. This work would have been impossible without you. I am also extremely grateful to Don Meyer who permitted me to post on the listserve.

References

Alzheimer's Association LA and Riverside. 2002. *Long Distance Caregiver Project*. [Online: Alzheimer's Association LA and Riverside] Available at: http://web.grcc.edu/ekunnen/crn/materials/LONGDISTANCECAREGIVER. html [accessed: 9 December 2007].

American Association on Intellectual and Developmental Disabilities (AAIDD). 2009. *FAQ on Intellectual Disability*. Online American Association on Intellectual and Developmental Disabilities (AAIDD). Available at: http:// aaidd.org/content_104.cfm [accessed: 31 July 2009].

Bigby, C. 1997. Parental substitutes? The role of siblings in the lives of older people with intellectual disabilities. *Journal of Gerontological Social Work*, 29(1), 3-21.

Butler, R. 2001. From where I write: the place of positionality in qualitative writing, in *Qualitative Methodologies for Geographers: Issues and Debates*, edited by M. Limb and C. Dwyer. London: Arnold Publishers, 264-276.

Dear, M.J. and Wolch, J.R. 1989. How territory shapes social life, in *The Power of Geography: How Territory Shapes Social Life*, edited by J.R. Wolch and M.J. Dear. Boston: Unwin Hyman, 1-18.

Dykens, E. 2005. Siblings: happiness, well-being, and character. *Mental Retardation*, 43(5), 360-364.

Eriksen, S. and Gerstel, N. 2002. A labor of love or love itself: care work among adult brothers and sisters. *Journal of Family Issues*, 23(7), 836-856.

Ferguson, P. and Asch, A. 1989. Lessons from life: personal and parental perspectives on school, childhood, and disability, in *Schooling and Disability*, edited by D. Biklen,, D. Ferguson and A. Ford. Chicago: The National Society for the Study of Education, 108-140.

Flaton, R. 2006. 'Who would I be without Danny?' Phenomenological case study of an adult sibling. *Mental Retardation*, 44(2), 135-144.

Goodchild, M. 2000. Topology, in *The Dictionary of Human Geography (4th edition)*, edited by R.J. Johnston et al. Oxford: Blackwell Publishers, 840.

Gorelick, J. 1996. *A Strong Sibling Network: Forgotten Child No More*, 10th Annual World Congress of the International Association for the Scientific Study of Intellectual Disabilities, Helsinki, Finland, 8-13 July 1996.

Grant, G. 2007. Invisible contributions in families with children and adults with intellectual disabilities. *Canadian Journal on Aging*, 26(1), 15-26.

Greenberg, J.S., Seltzer, M.M., Orsmond, G.I. and Krauss, M.W. 1999. Siblings of adults with mental illness or mental retardation: current involvement and expectation of future caregiving. *Psychiatric Issues*, 50(9), 1214-1219.

Hallman, B. and Joseph, A. 1997. Exploring distance and caregiver gender effects in eldercare: towards a geography of family caregiving. *The Great Lakes Geographer*, 4(2), 15-29.

Health Insurance Portability and Accountability Act (HIPAA) of 1996. U.S. Public Law 104-191.

Heller, T., Caldwell, J. and Factor, A. 2007. Aging family caregivers: policies and practices. *Mental Retardation and Developmental Disabilities Research Review*, 13, 136-147.

Hodapp, R., Glidden, L.M. and Kaiser, A. 2005. Siblings of persons with disabilities: toward a research agenda. *Mental Retardation*, 43(5), 334-338.

Huberman, A.M. and Miles, M.B. 1998. Data management and analysis methods, in *Collecting and Interpreting Qualitative Materials*, edited by N.K. Denzin and Y.S. Lincoln. Thousand Oaks, CA: Sage Publications, 179-210.

Joseph, A. and Hallman, B. 1996. Caught in the triangle: the influence of home, work and elder location on work-family balance. *Canadian Journal on Aging*, 15(3), 393-412.

Joseph, A. and Hallman, B. 1998. Over the hill and far away: distance as a barrier to the provision of assistance to elderly relatives. *Social Science and Medicine*, 46(6), 631-639.

Koerin, B.B. and Harrigan, M.P. 2002. P.S. I love you: long-distance caregiving. *Journal of Gerontological Social Work*, 40(1/2), 63-81.

Krauss, M.W., Seltzer, M.M., Gordon, R. and Friedman, D.H. 1996. Binding ties: the roles of adult siblings of persons with mental retardation. *Mental Retardation*, 34(2), 83-93.

Lawton, L., Silverstein, M. and Bengston, V. 1994. Affection, social contact, and geographic distance between adult children and their parents. *Journal of Marriage and Family*, 56 (February), 57-68.

MetLife Mature Market Institute. July 2004. *Miles Away: The MetLife Study of Long-distance Caregiving*. Findings from a National Study by the National Alliance for Caregiving with Zogby International [Online: MetLife Mature Market Institute]. Available at: http://www.maturemarketinstitute.com [accessed: 10 December 2007].

Milligan, C. 2000. 'Bearing the burden': towards a restructured geography of caring. *Area*, 32(1), 49-58.

Milligan, C. 2001. *Geographies of Care: Space, Place and the Voluntary Sector.* Hampshire, England: Ashgate.

Moss, P. 1999. Autobiographical notes on chronic illness, in *Mind and Body Spaces*, edited by R. Butler and H. Parr. London: Routledge, 155-166.

Parr, H. 2003. Medical geography: care and caring. *Progress in Human Geography*, 27(2), 212-221.

Parr, H. and Philo, C. 2003. Rural mental health and social geographies of caregiving. *Social and Cultural Geography*, 4(4), 471-488.

Rimmerman, A. and Raif, R. 2001. Involvement with and role perception toward an adult sibling with and without mental retardation. *Journal of Rehabilitation*, 67(2), 10-15.

Rizzolo, M.C., Hemp, R., Braddock, D. and Pomerantz-Essley, A. 2004. *The State of the States in Developmental Disabilities*. Washington, DC: American Association on Mental Retardation.

Roff, L.L., Martin, S.S., Jennings, L.K., Parker, M.W. and Harman, D.K. 2007. Long-distance parental caregivers' experiences with siblings. *Qualitative Social Work*, 6(3), 315-334.

Schoonover, C.B., Brody, E.M., Hoffman, C. and Kleban, M.H. 1988. Parent care and geographically distant children. *Research on Aging*, 10(4), 472-492.

Seltzer, G.B., Begun, A., Seltzer, M.M. and Krauss, M.W. 1991. Adults with mental retardation and their aging mothers: impacts of siblings. *Family Relations*, 40, 310-317.

Seltzer, M.M., Greenberg, J.S, Krauss, M.W., Gordon, R.M. and Judge, K. 1997. Siblings of adults with mental retardation or mental illness: effects on lifestyle and psychological well-being. *Family Relations*, 46(4), 395-405.

Seltzer, M.M., Greenberg, J.S., Orsmond, G.I. and Lounds, J. 2005. Life course studies of siblings of individuals with developmental disabilities. *Mental Retardation*, 43(5), 354-359.

Stoneman, Z. 2005. Siblings of children with disabilities: research themes. *Mental Retardation*, 43(5), 339-350.

Swenson, S. 2005. Families, research, and systems change. *Mental Retardation*, 43(5), 365-368.

Wiles, J. 2003. Daily geographies of caregivers: mobility, routine, scale. *Social Science and Medicine*, 57, 1307-1325.

Zeitlin, A.G. 1986. Mentally retarded adults and their siblings. *American Journal on Mental Deficiency*, 91, 217-225.

Chapter 8

Young People with Socio-Emotional Differences: Theorising Disability and Destabilising Socio-Emotional Norms

Louise Holt

Introduction

This chapter draws upon qualitative research with young people defined as having Special Educational Needs (SEN) and educational professionals, within mainstream secondary schools. A reconceptualisation of disability is offered. It is contended that current social model accounts of disability should be re-appraised to include the experiences of young people with socio-emotional differences, often defined within the UK education institution as having Emotional and Behavioural Difficulties (EBD).[1] I suggest that theorising the experiences of such young people under the lens of critical disability studies can more effectively conceptualise and politicise their experiences, challenging the dominant tendency within educational accounts to reproduce individual tragedy models of EBD. Further, analysing the experiences of such young people within critical disability studies can enhance theorisations of disability, by suggesting a need to more fully decouple disability from impairment and by pointing to a more expansive disability politics.

Three key theoretical tenets are drawn upon and expanded within this chapter. First, the widely accepted proposition that disability is not entirely 'natural', which raises the potential of more fully disentangling disability from impairment. This also facilitates developing more affirmative interpretations of disability, although such

---

1   The young people focused upon in this paper are those diagnosed with Social, Emotional and Behavioural Differences within the UK SEN institution. This definition is approximately equivalent to 'Emotional and Behavioural Disorders or Severe Emotional Disturbance' (Kauffman 2001) within the US context. This educational definition has overlaps with those in the International Classification of Disease, Functioning and Disability (WHO 2004). Such diagnoses are fluid. For instance, many young people with EBD also have learning disabilities or bodily differences. Frequently, young people diagnosed with EBD under the educational model of disability (Holt 2004a) do not have specific psychological or physiological diagnoses. However, young people diagnosed with EBD have specific diagnoses and become labelled with Attention Deficit and Hyperactivity Disorder (ADHD) or Autistic Spectrum Disorders, for instance. Many young people diagnosed as having EBD are not subject to such specific diagnoses, however.

a focus is not developed within this particular chapter. Second, given (dis)ability is increasingly understood non-dualistically (Thomas and Corker 2002, Imrie 1996, Chouinard 1997), the empirical focus of disability studies could be shifted to more fully incorporate individuals without specific medical or psychological diagnoses who are disabled via general and institutional socio-spatial practices. Third, recent attempts have been made to theorise embodiment within disability studies, by challenging the tendency to represent (impaired) bodies as a static pre-given. This challenge points to the need for more attention to be paid to how bodies' dynamic corporeality and socio-cultural/spatial configuration intersect to (re)produce diverse experiences of disability and disablement. The increasing emphasis on various bodies suggests a need to call for more attentiveness to a fuller range of mind-body-emotional differences than is currently explored.

The chapter has four key sections. First, the conceptualisation of disability is outlined, and the need to more fully incorporate socio-emotional differences is established. Second, brief details about the empirical study from which findings are drawn are presented. The third section traces some examples of how disability can be practiced around socio-emotional difference. The final section offers a concluding discussion.

(Re)theorising Disability – Incorporating Emotional Differences

Dynamic, Embodied Conceptions of (Dis)ability

In this chapter, disability is conceptualised as a dynamic, social and corporeal embodied identifier (Holt 2007, 2008). Bodies are inscribed with powerful socio-cultural categories (such as 'disabled'/'non-disabled') which entwine with corporealities or material bodies to reproduce hierarchical subject positionings. Ultimately, such embodied identities are key to the reproduction or transformation of broader socio-spatial (in)equalities.

The concept of disability employed takes as its departure point critical achievements of social model approaches (Oliver 1996a). These destabilise the location (and cause) of disability away from the individual body-mind towards disabling socio-spatial frameworks. However, some critiques levelled at social models are taken forward; albeit retaining the central insight of the socio-spatial production of disability. First, the chapter learns from critiques of social models as relatively disembodied (Crow 1996). Thus, here, disability is understood as a property of both dynamic, material bodies, and broader socio-spatial processes (Parr and Butler 1999). Moss and Dyck (1996) contend that it is often impossible to unravel the relative importance of the material over the social body. Tremain (2002) drawing upon Foucault, suggests that what is taken as the biological or material body (impairment) is already framed within powerful discourses of naming and hierarchical labelling. Thus, impairment is no more pre-social than disability, and the disability/impairment dichotomy becomes difficult, if not impossible, to maintain.

The perspective here is similar to Tremain's Foucauldian standpoint, and utilises a critical synthesis of Judith Butler's theories of performativity (Butler 1990, 1993) and subjection (Butler 1997, 2004) and Bourdieu's habitus and capitals (e.g., Bourdieu 1984, 1986, Bourdieu and Thompson 1991). It is contended that disability can be understood as a relational, embodied, corporeal and social identity positioning, which is (re)produced via a variety of interconnected institutional and non-institutional social and cultural mechanisms, including the everyday practices of reiteration of 'appropriate' performances (see Holt 2007, 2008). The construction of (dis)ability intersects with 'other axes of power relations' (Butler 1990: 4) to produce apparently fixed individual and natural embodied identities. These individual embodiments are a key mechanism for the reproduction or potential transformation of broader scale material inequalities.

It is increasingly accepted within critical social studies that (dis)ability should be understood non-dualistically (Corker 1999, Williams 2001, Hughes and Peterson 1997); disability is not an essential category (Disability). Individuals' experiences of disability intersect with a host of identifiers, including impairment type. However, this contention has not been fully empirically tested. Studies of disablement often focus upon the experiences of individuals with identifiable medical or psychological diagnoses (although for exceptions see Dorn 1999, Wilton 2007). This tendency can be traced to the retention of impairment as a necessary element of disability, which itself is often left unchallenged and under-theorised (Tremain 2002, Hughes and Peterson 1997, Shakespeare 1996). Ultimately, tying disability to impairment reproduces the biologism of the medical model. In taking medical or individual tragedy models as the starting point and seeking to oppose them, social model approaches have inadvertently reproduced some elements of these accounts.

Thus, the disablement of individuals without recognised impairments is under-represented, including those with learning differences and mental ill-health (Hall and Kearns 2001, Holt 2003a, b, 2004a, b, Philo and Metzel 2005, Goodley and Van Hove 2005). To date, the experiences of those with emotional differences (Holt 2003a, 2004 a, b, 2007) have not been fully explored within critical, social model rooted, conceptions of disability. The marginalisation of young people with socio-emotional differences within schools and broader society has been reflected in their relative silence in debates about both disability and the critical understandings of inclusion (Cooper 2003).

Including Socio-Emotional Differences

Within critical disability studies, there has been a recent move towards more fully incorporating the emotions. However, to date this has been limited to exploring psycho-emotional consequences of the experience of disablement (Thomas 1999) and, indeed, of impairment (Reeve 2003). The emotional turn in social sciences (Bondi et al. 2005) has thus had only limited impact on critical disability studies.

The experiences of people who are disabled by contrast to norms of socio-emotional performance have been under-explored within critical disability studies. Molloy and Vasil (2002) is a rare exception. They suggest the need to incorporate the experiences of young people with Asperger's 'Syndrome' (and implicitly other socio-emotional disabilities) within critical accounts of disability. They argue:

> We have little knowledge of the personal understandings of the very children we are categorising or the social repercussions of being labelled for these children (Molloy and Vasil 2002: 668).

The relative silence of such young people's experiences is partly associated with the adult-centricism of social studies of disability, which has recently been challenged (Watson et al. 1999, Davis and Watson 2001, Staker and Connors 2007). The diagnoses of emotional differences are particularly associated with childhood and youth, having a relatively recent emergence (Molloy and Vasil 2002, Broome-Smith and Smith 2007). Diagnosis is frequently instigated in schools by a child's inability to conform to classroom norms and expectations (Department for Education and Skills 2001).

The absence of people diagnosed with EBD in critical disability studies both reflects and reinforces their marginalised position in a variety of forums. For instance, such young people are, paradoxically, the group most likely to be excluded from mainstream education (Farrell and Tsakalidou 1999) and those around which resistance to 'inclusion policies' are often mobilised (MacBeath et al. 2005). They form an 'absent presence' within debates about educational inclusion (Cooper 2003, Farrell 2001), and this is reflected in the field of critical disability studies. This absence leaves unquestioned dominant, individual tragedy conceptions, which embody the cause of EBD within the individual.

Expanding conceptions of disability to include young people with socio-emotional differences is important on two key grounds. First, it facilitates the politicisation of EBD as a disability. Second, conceptualising EBD as a disability challenges representations of emotional differences as embodied within individuals and pre-existing the context of the school. Although arguably still the dominant mode of interpreting disability, individual tragedy models of disability have been significantly challenged by social model approaches. Without doubt, social model approaches are transforming socio-legal contexts of disability (Disability Discrimination Act (DDA), 1995/2005), although clearly further progress is required. By contrast, it is demonstrated in this chapter that dominant representations of socio-emotional difference as an individual pathology that pre-exists the social context of the school remain unchallenged. It is important to highlight the socio-spatial processes that disable individuals with socio-emotional differences, constructing a pseudo-medical category of EBD. Conceptualising EBD as a disability facilitates this critique.

Methodologies

This chapter presents findings from an in-depth, qualitative study with 18 young people with mind-body-emotional differences, diagnosed with 'Special Educational Needs' (SEN) within the English (special) educational institution. The study aimed to explore young people's (age 12-16) experiences of social inclusion, exclusion and social capital within mainstream school and leisure spaces. The research comprised three repeated focus groups and a semi-participatory, visual technique with groups of young people from three secondary (high) schools.[2] The participatory visual methods were not analysed in isolation; rather they were used to inform discussions with individual children during group activities. Eight semi-structured interviews were also conducted with education actors in the Local Education Authority (LEA) and schools.

Efforts were made to ensure that young people consented rather than assented to the research. Discussions in classes about the research preceded giving the young people a letter with information detailing the research, with sections aimed at both the young people themselves and their parents/carers. The research participants were requested to sign a consent form themselves, and to bring a form signed by a parent/carer. Both the researcher and staff in the schools emphasised that participation in the research was voluntary, and that individuals could withdraw at any point, and did not have to take part in any activities or answer any questions with which they were uncomfortable.

The chapter explores the experiences of young people with a range of SEN diagnoses, whose behavioural practices conflict with school socio-emotional norms. The majority of participants had a diagnosis of EBD. However, some of the young people, with ostensibly similar behaviour and experiences, were not so labelled. Table 8.1 gives some details of the participants. All of the pupils are white and broadly 'working class' and/or their parents are unemployed. Andrew is also white, but his parents have professional jobs. The homogeneous ethnic/racial composition of the participants reflects the population profile of the case study location.

(Dis)abling Young People with Socio-Emotional Differences

In this section, I trace two key and interconnected facets of the (re)production of young people with socio-emotional differences as (dis)abled within school institutional spaces. First, the diagnosis of young people as disabled via the

---

2 The young people were given disposal cameras and requested to take photographs of an average school day and an average non-school day. The photographs were developed and handed back to the young people. The participants constructed caption storyboards using the photographs, and kept any photographs not used. Some young people drew pictures rather than taking photos. Young people annotated their pictures with either sentences or single words.

Table 8.1   Profiles of study participants

Individual	Diagnosis
Adam	15 year old boy diagnosed with specific learning differences and specific EBD
Amy	13 year old girl with chronic illness and behaviour differences – not defined as having EBD
Andrew	Boy with specific learning difficulties and some emotional differences (not diagnosed with EBD)
Annie	14-15 year old girl diagnosed with EBD
Claire	13 year old girl diagnosed with EBD
David	Boy with learning differences. Diagnosed with EBD
Helen	14 year old girl with visual impairments who is also a wheelchair user. Some behavioural differences, not diagnosed with EBD
Jade	13-14 year old girl with some body differences diagnosed with EBD
Kevin	13-14 year old boy diagnosed with EBD
Lee	13 year old boy diagnosed with EBD
Lewis	13 year old boy diagnosed with EBD
Michael	14-15 year old boy diagnosed with specific learning differences and EBD
Neil	14-15 year old boy diagnosed with ADHD and later re-diagnosed with ASD
Peter	14 year old boy with body differences, diagnosed with EBD
Tom	13-14 year old boy diagnosed with EBD

Note: All names are pseudonyms.

institutional powers and resources of the SEN institution is explored, through analysis of policy documents and interviews with educational professionals. Second, brief discussion is presented of how young people are (dis)abled through everyday socio-spatial practices of children and adults in school spaces. These two facets of disability often intertwine, and can be mutually reinforcing. However, they also occasionally diverge.

Pseudo-Medical Educational Diagnoses of EBD

Young people with socio-emotional differences can be diagnosed as having EBD via the 'powers and resources' (Philo and Parr 2000) of the special education institution, when their behaviour falls outside normative expectations of deportment within classroom settings, if their conduct is considered to inhibit their learning. Young people become identified as having SEN specifically when:

> Children have special educational needs if they have a *learning difficulty* which calls for *special educational provision* to be made for them.

Children have a *learning difficulty* if they:

> i) have a significantly greater difficulty in learning than the majority of children of the same age; or ii) have a disability which prevents or hinders the child from making use of educational facilities of a kind generally provided for children of the same age in schools within the area of the local education authority (Department for Education and Skills 2001, 1/3, original emphasis).

Requests for intervention are usually initiated by either classroom teachers or parents. Definitions of EBD are slippery (see Visser 2003, Cole and Visser 2005) and underpinned by the identification of non-normative behaviour that forms a significant barrier to learning. The most recent Code of Practice defines children with EBD, loosely, as those:

> ...who demonstrate features of emotional and behavioural difficulties, who are withdrawn or isolated, disruptive and disturbing, hyperactive and lack concentration; those with immature social skills; and those presenting challenging behaviours arising from other complex special needs... (Department for Education and Skills 2001: 7:60).

Definitions of EBD are rendered complex given a broader tendency to characterise behaviour that conflicts with societal norms as a defining component of youth (Valentine 1996). Indeed, dominant discourses increasingly demark all or most young people as exhibiting unruly and threatening behaviour (Manning et al. forthcoming, Lees 2003), yet only a in a minority of cases does the behaviour of young people lead to labelling with Special Educational Needs. EBD can therefore be viewed as 'pathological' levels of non-normative behaviour; when unruly behaviour becomes framed as an illness to be treated. It is a slippery concept given accepted and expected levels of 'disruptive' behaviour vary across educational spaces. It is therefore unsurprising that diagnoses of EBD are socio-spatially specific, and vary between schools even within the same city (Galloway et al. 1994, Kelly and Gray 2000, Holt 2004a). Despite recent attempts to pin-down and centralise norms of socio-emotional development (Kelly 2005), these expectations remain relatively hidden, and have not been subject to the same long-term formal scrutiny and measurement as the learning norms which underpin formal curricula and assessment. Kauffman (2001: 23) contends that:

> An emotional or behavioural disorder is whatever a culture's chosen authority figures designate as intolerable. Typically, it is that which is perceived to threaten the stability, security, or values of that society. Defining an emotional or behavioural disorder is unavoidably subjective, at least in part.

This statement bears parallels to Szasz's (1974) conception of psychiatry and mental ill-health in suggesting that socio-emotional differences are primarily social constructions (see also Cooper 2001). I would argue that there are also embodied, corporeal, experiences of social and emotional difference (see also Parr 1999). To deny this is to ignore the very real 'pain and confusion' that can be caused by these differences. However, the socio-cultural and spatially variable labels and practices that surround young people also frame their experiences, and emotional differences, although potentially embodied, do not pre-exist these social contexts. The socio-cultural and corporeal elements of these disabling differences cannot be unravelled; mind-body-emotions and society/space intersect in complex ways. It is, however, crucial to trace some of the causes of socio-emotional disablement in society and space, to redress the balance of searching for explanations within the psychology of the young person.

Slippery, Shifting Diagnoses

The pseudo-medical labels attached to young people with EBD can be either general or specific and tied to an often contested 'condition'. Frequently, the diagnoses applied to individual young people change between different categories of EBD and mental ill-health:

> Neil was initially diagnosed as having Attention Deficit and Hyperactivity Disorder (ADHD). He was recently re-diagnosed as being on the Autistic Spectrum…He's being having a lot of difficulties with inappropriate behaviour…
> (Ms. Woods, school 3).

The slippery nature of the diagnosis (Visser 2003, Cole and Visser 2005) is tied to the various powers and resources available to different pseudo-medical institutions. Of particular relevance here is the acutely under-resourced young people's mental health facilities in the UK context (Kelly and Gray 2000), reflecting broader patterns in mental health care provision (Wilton and Evans 2007).

There is both overlap and divergence between definitions of SEN within education institutions and 'disability' within the DDA (1995/2005) in the UK context. The DDA emphasises the presence of specific impairments or 'conditions' whereas the SEN institution is based upon difficulties in learning. The boundaries between the 'medical' and pseudo-medical are becoming increasingly blurred, given the enhanced medical intervention, such as medication, targeted at young people with specific EBD (Miller et al. 2001). Medical and educational diagnoses and experiences of mind-body-emotional differences are often complex and interconnecting. As the experiences of young people in this study testify (Table 8.1), individuals often have multiple mind-body-emotional differences and diagnoses (Department for Education and Skills 2001).

Individual Tragedy Models of EBD and Pathologising the Family

The majority of diagnoses to which young people with EBD are subject are individual, pathologising labels (Oliver 1996b) which reflect individual tragedy models of disability by being firmly stated as an internal psychological/educational condition (see also Finkelstein and French 1993). Once diagnosed with EBD, any problems that young people with socio-emotional differences experience with either formal or informal aspects of schools are firmly located within the individual, who is frequently defined as having 'poor social skills'. As the following excerpt emphasises:

> Ms. Brown: [Michael]…is a nice boy – he hangs around with the wrong characters sometimes, but he's…
>
> Ms. Woods: He's got social issues, hasn't he?
>
> Ms. Brown: He does struggle in terms of his relationships with other students (Joint interview with Ms. Brown and Ms. Woods, school 3).

The propensity of adults to reproduce individual tragedy models of disability by blaming disabled children for social and learning problems has been noted in relation to children with body and learning differences (Watson et al. 1999). However, this tendency is particularly marked for young people with EBD (see also Holt 2003a, 2007). The behavioural underpinnings of the diagnosis operate as a carte blanche to explain any difficulties with learning or peers, without seeking explanations within the socio-spatial context of the school.

In contrast to broader individual tragedy models of disability, social contexts are often recognised as a causal factor of EBD. Almost exclusively, however, the social context of EBD is viewed as the home and family background. This is viewed as a social context that has been concretised within the psyche of the individual with EBD and becomes a pre-existing, embodied fact:

> Annie is one of the most difficult students we have in the school, and she's statemented purely for behaviour. Nothing wrong with her literacy or anything. She's been passed from pillar to post…she is looked after by the Local Authority. I think she's in foster care. She does have some relationship with her sister. I doubt that she will turn up for the research (Ms. Woods, school 3).

Thus, the 'blaming' for socio-emotional differences can be broadened out beyond the individual towards the family. The disability remains a tragedy; an accident of birth, of having been born into a family whose deviance is embodied and concretised in an individually pathologised set of behaviours, which are seen to require pseudo-medical, psychological, intervention. Crucially, the social context of EBD (the family) is held to be solidified within individual children's bodies. The

EBD becomes embodied as a facet of the individual which can be accommodated within the school context, but which exists independently of the socio-cultural and spatial framework of the school. The EBD pre-exists the school environment. The school space is not viewed as a component of the EBD.

This tendency to locate the cause of the EBD within deviant families was generally not reflected in discussions about people who also had body differences. Rather than exploring social contexts that might have led to the emergence of EBD, practices that conflict with school norms are usually conceptualised as 'natural' emotional and psychological responses to a problematic body:

> She [Helen] can be incredibly rude. She can be very, very rude but I've always said to her that she has to come and tell me because it must be so frustrating, bless her. The only thing she's got is her mouth isn't it? She can't see body language, she can't get atmosphere so she shouts and she gets very frustrated with things and these VI people are very, very good but when you've got them with you all the time and you've got to have somebody with you all the time, you don't want somebody with you all the time, you want to go out with your mates and go round the school with your mates but you can't do that because people are afraid of what might happen so she gets very angry (Ms. Tyler, school 2).

This further defines people with body differences under the individual tragedy model as 'victims' whose psychological state is associated with their bodily difference (Finkelstein and French 1993). In a similar vein to debates about depression and even suicide among disabled people (Morris 1991) EBD is viewed as a natural response to the tragic individual circumstance of disability, and as a component of this individual pathology. This reinforces the discourse of disability as an individual random tragedy. Broader social explanations for EBD are also not sought for these children. This is problematic as this negates the possibility that the behaviour is rooted in problematic social experiences in school, home or other contexts.

The social context of the school rarely features as a factor within professional and policy discussions of the causes of EBD. When discussed, changes in school environments are viewed as adjustments to accommodate differences which arise from inherent characteristics of the young person which may have been initially caused by social contexts outside of the school but which are a-priori to the school context, and embodied within the individual young person:

> ...he [Ant] is finding the gap is widening and what we are hoping to do next year with him, he's one of the pupils, when he gets to Year 10, we will probably give him an alternative curriculum which means that we'll probably send him out for one day a week to somewhere like a farm, we have...or the nursery, where he can do more practical things, have him out for a whole day, and just be in school more or less for the core subjects and possibly not even do GCSEs but a Certificate of...Lifeskills, not a full curriculum by any means (Ms. Tyler, school 2).

In line with other individual tragedy responses to disability, changes in the socio-spatial environments for people with socio-emotional differences are presented as compensations for their internal 'different' characteristics. It is not generally explicitly acknowledged that EBD emerge at a dynamic interface between the mind-body and all the socio-spatial contexts through which young people's lives and identities are constructed, including the school.

The peripheral role of the school in teachers' and policy discussions of the emergence of socio-emotional differences is both surprising and problematic, as the school is a key site for the disablement of young people on socio-emotional grounds. The experience of EBD is statistically associated with social class, race/ethnicity and gender, with socially excluded, 'working class', males and certain ethnic minority groups over-represented among individuals defined as having EBD (Thacker et al. 2001). This link is unsurprising given that normative educational frameworks are bourgeois, racialised and gendered, suggesting that the 'cause' of EBD might be in the education institution rather than homes (Bourdieu and Passeron 1979, Khattab 2005, Modood 2003). The majority of the young people defined as having EBD in this study were working class or socially excluded. This suggests that EBD can be a mechanism for the reproduction of socio-economic exclusion. However, further research is required to tease-out how EBD reproduces socio-economic exclusion and how these patterns can be transformed (see also Slee 2006).

(Dis)abling Young People with Socio-Emotional Differences in Formal Education Spaces

Within-School Exclusions and Segregations

The macro-scale pattern of high levels of from-school exclusion of young people with socio-emotional difference was reproduced at the micro-scale of the study schools, despite their avowedly inclusive policies. In all of the study schools, students with socio-emotional differences were frequently segregated from mainstream lessons or social activities. These segregations and exclusions ranged from: exclusions from playground spaces as a punishment for inappropriate behaviour and/or for protection from bullies; removal from mainstream classrooms for short periods as a punishment, and more systematically to follow alternative curricula; and attending separate 'units' or following an alternative curriculum.

In one study school, some students spent a proportion of their time in a 'special facility' within the school away from the other students. Students were selected to spend some time within the 'special facility' on the grounds that they were not benefiting fully from the mainstream curriculum, usually due to exhibiting behaviour which conflicted with school norms. Most of the individuals who were thus segregated were seen to have complex issues:

...there were a lot of issues in society that were interfering with students' ability access the curriculum for whatever reason, you know, home issues – some were in care, some of them were travelling, some of them had dyslexia which was causing frustration. A whole host of kids where, you know, the mainstream curriculum day in, day out was not working (Ms. Brown, school 3).

This micro-institution was characterised by a relatively high ratio of non-teaching staff to students, the provision of youth workers, more relaxed rules, and some therapeutic facilities. Despite the broader scale exclusion involved, many young people valued this space:

LH: What about you, what do you like about school?

Annie: It's all crap.

LH: Why's that?

Annie: Well, [the special facility] is better, but the lessons – you don't fucking learn anything – all you hear are the teacher's moaning. In [the special facility] they are less stressy...less stupid rules (Annie, school 3, focus group 2).

That young people often felt more comfortable within such segregated spaces questions simplistic representations of 'exclusion' as reproducing difference negatively, and points to the need to consider the nuanced experiences of exclusion/segregation (see also Parr 1997, Holt 2004b, Hall 2005). The agency behind the decision to include/exclude individuals from/within different settings and/or the preferences of the young people who are segregated is of importance (Cook et al. 2001). However, emphasising young people's preferences could problematically suggest that they are sovereign agents rather than partially knowledgeable subject/agents. All individuals' practices are simultaneously conscious and unconscious and often unreflectively reproduce dominant social relations within which they are themselves positioned (Holt 2008). Thus, even if young people would chose to be segregated, such patterns of segregation map 'otherness' on the school space and are a mechanism for reproducing difference negatively. Teachers also emphasised that, paradoxically, the within-school segregation facilitated a broader scale school inclusion, preventing or delaying the exclusion of young people from mainstream schools.

Contrasting to Behavioural Norms, and the Role of Normalising Therapies

Young people are typically diagnosed with EBD on the grounds that they are unable to conform to expectations of behaviour within school spaces. Thus, being (dis)abled by contrast to school norms is integral to socio-emotional differences. It is therefore not surprising that these young people are often subject to increased surveillance in an endeavour to normalise their behaviour. This enhanced surveillance lead to young people with socio-emotional differences often being the targets for punishment, even for acts that they did not commit. As Peter (school 2) elucidates:

> Apparently I was up and I just smashed…trashed an old school…On August 20th this year. Apparently I trashed that school, smashed about five windows and I've now got caught five times…I weren't even there. Someone just…popped my name out in their heads…

Individuals with EBD are also (dis)abled by contrast to norms of socio-emotional deportment in classroom spaces that are reproduced by both teachers and other young people. The participants gave many examples of their practices that conflicted with expectations within specific spaces. Incidents discussed ranged from being 'cheeky' or 'talking back' to violent or aggressive acts. For example:

> She – I wasn't very happy that day and so – 'cos I used to then go under tables and she went and grabbed me didn't she? And I just went so I just went – and bit her, and got kicked out of school for a little while and I had to go and see this psych person (Jade, school 1, focus group 2).

As the discussion above emphasises, young people with socio-emotional differences are often subject to normalising 'therapies' aimed at treating their individual deviance. Social models of EBD suggest that such normalisation is problematic, as it is underpinned by an assumption that the children with socio-emotional differences need to change, and fails to more fully tease out the interface between socio-spatial processes and individual bodies that is the site of the (re)production of EBD (see also Copeland 1999). However, the young people in this focus group went on to compare notes about their shared, and often positive, experiences of counselling. Lewis states:

> Lewis: I see a one of them people – a psychologist. Not the peoples who lie you down on the bed – it's the people who talk to you about your problems and find out about your problems and all that.
>
> LH: And do you like doing that, do you like going to see those people? Or do you not like it?
>
> Lewis: It's alright, to find out why I do some things and why it happens sometimes and all that (Lewis, school 1, focus group 3).

Punishment, Discipline and Anti-School Norms

Conflicting with school behavioural expectations leads to young people frequently being subject to high levels of punishment, which is a disciplinary strategy drawn upon for the normalisation of deviant students. The research participants discussed their experience of punishment.

> Helen: I had detention yesterday well funny.
>
> Peter: Did you go?
>
> Helen: Yeah!
>
> Amy: What a loser!
>
> Helen: Do you know what, right I got a double detention for disruption.
>
> Amy: I got a detention for trying to get out of PE then when I said something, I give it away!
>
> Lee: [Says quietly and just audibly for the tape] Psychos (school 2, focus group 2).

These punishments seem unlikely to turn these young people into conforming citizens. Rather, the research participants viewed these disciplinary strategies as a badge of honour. Hence, the young people re-signify the meaning of these 'punishments'. The focus group context becomes a new arena for playing out identities in specific ways; there is an attempt by the young people to out-compete their peers in rule breaking behaviour and punishments received. Although selectively emphasising their own rule-breaking practices, these discussions highlight the anti-school norms reproduced by many of the young people who participated in the study. Thus, here, young people with socio-emotional differences can be particularly *included* within specific sub-cultural groups (although there are limits to such inclusion, as Lee's comment suggests).

Discussion and Conclusion

In this chapter, I have argued that EBD should be conceptualised as disabilities, within the framework of critical, social model inspired theories. The under-representation of people with socio-emotional differences within critical disability studies is problematic. Although their experiences can be aligned with those of other groups with mind-body differences subject to dominant, individual tragedy models of disability, this group has been excluded from the wider political re-signification of disability. This re-signification has seen a move away from individual pathology representations and practices, towards critical understandings of disability as situated either primarily in society and space, or simultaneously in society-space and bodies. Of course, this transformation is by no means complete, and dominant, negative conceptions of disability are difficult to transform.

Young people with socio-emotional differences experience entrenched disablism. The lack of a critical, social model inspired account of EBD with which to challenge dominant perspectives is one factor underpinning the relative marginalisation of young people with these types of mind-body-emotional differences. As a starting point to building a critical account of the experiences of young people with socio-emotional differences, some elements of such young people's disablement within the socio-spatial context of schools have been traced in this chapter. Young people with socio-emotional differences are disabled by contrast to behavioural norms and subject to a variety of disabling practices, such as (although not limited to) labelling and diagnostic processes, and various exclusions/marginalisations within and from school institutional spaces. Thus, the disability of such young people is located within the social and spatial contexts through which they live their lives at least as much as within their individual mind-bodies (which themselves are not bounded, but become in relation to society and space). It is suggested that this re-signification ties into, and points to a need for, a more inclusive politics of disability which allows some of the advances made by disabled people and their allies to be expanded to other 'other' groups (those marginalised within both broader society and disability politics and culture).

The experience of young people who contravene normative expectations of behaviour exposes the broader societal norms, which are reproduced in dynamic, shifting ways in specific school spaces. These norms usually remain hidden and are relatively unproblematically embodied in the majority of individuals, albeit in slightly diverse ways that contain the potential for transformation. However, it would be problematic to somehow celebrate the young people's disruption of these social norms as an act of active resistance to broader society. Society is governed by these norms, and a failure to embody them leads to interconnected social exclusions, both currently and in the future. Clearly, societies require norms and regulations to function. As Katz (2004) states, suggesting that social categories are constructions should be a starting point, rather than an end point. By contrast, it might be more pertinent, albeit more modest, to question who is advantaged by these norms and who is (dis)abled? How do these norms reproduce gendered, sexualised, classed, racialised and (dis)abled subjects, and with what material effects?

On a more practical level, this chapter has pointed to some features of what more inclusive schools might look like. Some of the segregated spaces of the 'special unit' and therapeutic spaces were viewed by the research participants as relatively inclusive. These have resonance for broader suggestions for constructing more inclusive schools, with smaller classes, less hierarchical relationships between adults and young people and space to discuss feelings. To become inclusive to all young people, the entire education institution needs to be radically altered, given the transformation in the distribution of powers and resources that the development of inclusive schools would require. Normative expectations of behaviour and ability are central underpinnings of the education institution which would also require radical transformation. However schools are not isolated, but are components of broader society, wherein broader normative values are (often subconsciously) reproduced. This renders transformation difficult, although not impossible to achieve.

References

Bondi, L., Davidson, J. and Smith, M. 2005. Introduction: geography's 'emotional turn', in *Emotional Geographies*, edited by J. Davidson, M. Smith and L. Bondi. London: Ashgate.

Booth, T. and Ainscow, M. 2002. *Index for Inclusion*. Bristol: Centre for Studies in Inclusive Education.

Bourdieu, P. 1984. *Distinction: A Social Critique of the Judgement of Taste*. London/New York: Routledge.

Bourdieu, P. 1986. The forms of capital, in *Handbook of Theory and Research in the Sociology of Education*, edited by J.G. Richardson. New York: Greenwald Press.

Bourdieu, P. and Passeron, J.C. 1979. *The Inheritors: French Students and their Relation to Culture*. Chicago: University of Chicago Press (tr. Nice, R.).

Bourdieu, P. and Thompson, J.B. 1991. *Language and Symbolic Power*. Cambridge MA: Harvard University Press.

Broome-Smith, J. and Smith, D. 2007. *The Geographic Complexities of Autistic Spectrum 'Disorders': A Local Perspective?* Paper presented at the First International Conference on Geographies of Children, Youth and Families, University of Reading, 17-18 September.

Butler, J. 1990. *Gender Trouble: Feminism and the Subversion of Identity*. London/New York: Routledge.

Butler, J. 1993. *Bodies that Matter: On the Discursive Limits of 'Sex'*. London/New York: Routledge.

Butler, J. 1997. *The Psychic Life of Power: Theories in Subjection*. Stanford: Stanford University Press.

Butler, J. 2004. *Undoing Gender*. London/New York: Routledge.

Chouinard, V. 1997. Making space for disabling difference: challenging ableist geographies. *Environment and Planning D: Society and Space*, 15, 379-387.

Cole, T. and Visser, J. 2005. *Review of Literature on EBD Definitions and Good Practice Accompanying the Managing of Challenging Behaviour*. London: OFSTED.

Cook, T., Swain, J. and French, S. 2001. Voices from segregated schooling: towards an inclusive education system. *Disability and Society*, 16, 290-310.

Cooper, P. 2001. *We Can Work It Out: What Works in Educating Pupils with Social, Emotional and Behavioural Difficulties Outside Mainstream Classrooms?* Barkingside: Barnardo's.

Cooper, P. 2003. Editorial: including students with social, emotional and behavioural difficulties in mainstream secondary schools. *Emotional and Behavioural Difficulties*, 8, 5-6.

Copeland, I.C. 1999. Normalisation: an analysis of aspects of special educational needs. *Educational Studies*, 25, 99-111.

Corker, M. 1999. Differences, conflations and foundations: the limits to 'accurate' theoretical representation of disabled people's experience? *Disability and Society*, 14, 627-642.

Crow, L. 1996. Including all our lives: renewing the social model of disability, in *Encounters with Strangers: Feminism and Disability*, edited by J. Morris. London: The Women's Press.

Davis, J.M. and Watson, D. 2001. Where are the children's experiences? Cultural and social exclusion in 'special' and 'mainstream' schools, *Disability and Society*, 16, 671-687.

Department for Education and Skills. 2001. *Special Educational Needs Code of Practice.* London: Department for Education and Skills.

Disability Discrimination Act 2005. London: The Stationery Office.

Dorn, M. 1999. The moral topography of intemperance, in *Mind and Body Spaces: Geographies of Illness, Impairment and Disability*, edited by R. Butler and H. Parr. London/New York: Routledge.

Farrell, P. 2001. Special education in the last twenty years: have things really got better? *British Journal of Special Education*, 28, 3-9.

Farrell, P. and Takalidou, K. 1999. Recent trends in the re-integration of pupils with emotional and behavioural difficulties in the United Kingdom. *School Psychology International*, 20, 323-337.

Finkelstein, V. and French, S. 1993. Towards a psychology of disability, in *Disabling Barriers – Enabling Environments*, edited by J. Swain, V. Finkelstein, S. French and M. Oliver. London: Open University/Sage Publications.

Foucault, M. 1982. Subject and power, in *Michel Foucault: Beyond Structuralism and Hermeneutics*, edited by H. Dreyfus and P. Rabinow. Chicago, IL: University of Chicago Press.

Foucault, M. 1988. Technologies of the self, in *Technologies of the Self: A Seminar with Michel Foucault*, edited by L.H. Martin, H. Gutman and P.H. Hutton. Amherst, MA: University of Massachusetts Press.

Foucault, M. 1990. *The History of Sexuality, Vol. 1, An Introduction.* New York: Vintage Books.

Galloway, D., Armstrong, P. and Tomlinson, S. 1994. *The Assessment of Special Educational Needs.* London: Longman.

Goodley, D. 2004. Who is disabled? *Exploring the scope of the social model of disability, Disabling Barriers – Enabling Environments*, edited by J. Swain, V. Finkelstein and M. Oliver. London: Sage.

Goodley, D. and Van Hove, G. (eds) 2005. *Another Disability Studies Reader: People with Learning Difficulties and a Disabling World*. Antwerp: Garant.

Gregory, D. 1994. *Geographical Imaginations*. Oxford: Blackwell.

Hall, E. 2005. The entangled geographies of social exclusion/inclusion for people with learning disabilities. *Health and Place*, 11, 107-15.

Hall, E. and Kearns, R. 2001. Making space for the 'intellectual' in geographies of disability. *Health and Place*, 7, 237-246.

Holt, L. 2003a. *Disabling Children in Primary School Spaces*. Unpublished PhD thesis, Loughborough, Loughborough University.

Holt, L. 2003b. (Dis)abling children in primary school spaces – geographies of inclusion and exclusion. *Health and Place*, 9, 119-128.

Holt, L. 2004a. Childhood disability and ability: disablist geographies of mainstream primary schools. *Disability Studies Quarterly*, 24(3), unpaginated.

Holt, L. 2004b. Children with mind-body differences: performing (dis)ability in primary schools classrooms. *Children's Geographies*, 2, 219-36.

Holt, L. 2006. Embodying and destabilising (dis)ability and childhood, in *Contested Bodies of Childhood and Youth*, edited by K. Hörschelmann and R. Colls. New York: Palgrave.

Holt, L. 2007. Children's socio-spatial (re)production of disability in primary school playgrounds. *Environment and Planning D: Society and Space*, 25, 783-802.

Holt, L. 2008. Embodied social capital and geographic perspectives: performing the habitus. *Progress in Human Geography*, 32, 227-246.

Hughes, B. and Peterson, K. 1997. The social model of disability and the disappearing body: towards a sociology of impairment. *Disability and Society*, 12, 325-340.

Imrie, R. 1996. Ableist geographies, disablist spaces: towards a reconstruction of Golledge's geography and the disabled. *Transactions of the Institute of British Geographers*, 21, 397-403.

Katz, C. 2004. *Growing up Global: Economic Restructuring and Children's Lives*. Minneapolis: University of Minnesota Press.

Kauffman, J.K. 2001. *Characteristics of Emotional and Behavioural Disorders of Children and Youth*. New Jersey: Prentice Hall. 7th Edition.

Kelly, D. and Gray, C. 2000. *Educational Psychology Services (England): Current Role, Good Practice and Future Directions*. London: Department for Education and Employment.

Kelly, R. 2005. *Improving Behaviour in Schools*. London: Department for Education and Skills.

Khattab, N. 2005. Inequality and polarities in educational achievement amongst Britain's ethnic minorities. Leverhulme Programme on Migration and Citizenship Working Paper.

Lees, L. 2003. The ambivalence of diversity and the politics of urban renaissance: the case of youth in downtown Portland, Maine. *International Journal of Urban and Regional Research*, 27, 613-34.

Macbeath, J., Galton, M., Steward, S., Macbeath A. and Page, C. 2006. *The Costs of Inclusion*. Cambridge: University of Cambridge Press.

Manning, R., Jago, R. and Fionda, J. 2010. Socio-spatial experiences of young people under anti-social behaviour legislation in England and Wales, in *Geographies of Children, Youth and Families: International Perspectives*, edited by L. Holt. London: Routledge.

Miller, A.R., McGrail, K.M. and Armstrong, R.W. 2001. Prescription of methylphenidate to children and youth, 1990-1996. *Canadian Medical Association Journal*, 165, 1489-1494.

Modood, T. 2003. Ethnic differentials in educational performance, in *Explaining Ethnic Differences*, edited by D. Mason. London: ESRC and The Policy Press.

Molloy, H. and Vasil, L. 2002. The social construction of Asperger's syndrome: the pathologising of difference? *Disability and Society*, 17, 659-669.

Morris, J. 1991. *Pride Against Prejudice: Transforming Attitudes to Disability*. London: The Women's Press.

Moss, P. and Dyck, I. 1996. Inquiry into environment and body: women, work and chronic illness. *Environment and Planning D: Society and Space*, 14, 737-757.

Ofsted 1999. *Principles into Practice: Effective Education for Pupils with Emotional and Behavioural Difficulties*. London: Ofsted.

Oliver, M. 1996a. *Understanding Disability: From Theory To Practice*. Basingstoke: Macmillan.

Oliver, M. 1996b. Defining impairment and disability: issues at stake, in *Exploring the Divide*, edited by C. Barnes and G. Mercer. Leeds: The Disability Press.

Parr, H. 1997. Mental health, public space and the city: questions of individual and collective access. *Environment and Planning D: Society and Space*, 15, 435-454.

Parr, H. 1999. Delusional geographies: the experiential worlds of people during madness/illness. *Environment and Planning D: Society and Space*, 17, 673-690.

Parr, H. and Butler, R. 1999. New geographies of illness, impairment and disability, in *Mind and Body Spaces: Geographies of Illness, Impairment and Disability*, edited by R. Butler and H. Parr. London: Routledge.

Parr, H. and Philo, C. 1995. Mapping mad identities, in *Mapping the Subject*, edited by S. Pile and N. Thrift. London: Routledge.

Philo, C. and Metzel, D. 2005. Introduction to theme session geographies of intellectual disability: 'outside the participatory mainstream?' *Health and Place*, 11, 77-85.

Reeve, D. 2003. Negotiating psycho-emotional on disability and their influence on identity constructions. *Disability and Society*, 15, 492-508.

Shakespeare, T. 1996. Disability, identity and difference, in *Exploring the Divide*, edited by C. Barnes and G. Mercer. Leeds: The Disability Press.

Slee, R. 2006. Limits to and possibilities for educational reform. *International Journal of Inclusive Education*, 10, 101-110.

Staker, K. and Connors, C. 2007. Children's experiences of disability: pointers to a social model of childhood disability. *Disability and Society*, 22, 19-32.

Szasz, T. 1974. (1961). *The Myth of Mental Illness: Foundations of a Theory of Personal Conduct*. New York: Harper and Row.

Thacker, J., Babbage, E. and Strudwick, D. 2001. *Educating Children with Emotional and Behavioural Difficulties: Inclusive Practice in Mainstream Schools*. London/New York: Routledge.

Thomas, C. 1999. Narrative identity and the disabled self, in *Disability Discourse*, edited by M. Corker and S. French. Buckingham: Open University Press.

Thomas, C. and Corker, M. 2002. A journey around the social model, in *Disability/ Postmodernity*, edited by M. Corker and T. Shakespeare. London/New York: Continuum.

Tremain, S. 2002. On the subject of impairment, in *Disability/Postmodernity*, edited by M. Corker and T. Shakespeare. London: Continuum.

Valentine, G. 1996. Angels and devils: moral landscapes of childhood. *Environment and Planning D: Society and Space*, 14, 581-99.

Visser, J. 2003. *A Study of Children and Young People who Present Challenging Behaviour*. Birmingham, University of Birmingham.

Watson, N., Shakespeare, T., Cunningham-Burley, S., Barnes, C., Corker, M., Davis, J. and Priestley, M. 1999. *Life as a Disabled Child: A Qualitative Study of Young Disabled People's Perspectives and Experiences*. ESRC End of Project Report, University of Leeds.

Williams, G. 2001. Theorising disability, in *International Handbook of Disability Studies*, edited by G. Albrecht, K. Seelman and M. Bury. London: Sage.

Wilton, R. 2007. *The Politics of Place in Addiction Treatment*, paper presented at the Annual Meeting of the Association of American Geographers, San Francisco, 20 April.

Wilton, R. and Evans, J. 2007. *Beyond Landscapes of Despair Panel Session*, Annual Meeting of the Association of American Geographers, San Francisco, 17 April.

World Health Organization. 2004. *International Classification of Disease, Functioning and Disability*. Geneva: WHO.

Zola, I. 1991. Bringing our bodies and our selves back in: reflections on a past, present and future medical sociology. *Journal of Health and Social Behaviour*, 32, 1-16.

Chapter 9

Evaluating Workfare:
Disability, Policy and the Role of Geography

Claire Edwards

Introduction

The role that geographers and geographical analysis can play in shaping and evaluating public policy has become a renewed focus of debate within the academy (Dorling and Shaw 2002, Martin 2001, Pain 2006). This is particularly so in a context where the current United Kingdom (UK) government has placed an emphasis on the notion of evidence-based policy, which in and of itself has spawned a huge industry of commissioned research reports and policy evaluations about 'what works' (Sanderson 2003, Walker 2001, Davies, Nutley and Smith 2000). The evidence-based agenda has provoked debate amongst geographers and other social scientists about a number of issues, including what constitutes evidence, the relationship between (geographical) research and policy, and the epistemological underpinnings of the evidence-based enterprise itself (Imrie 2004, Sanderson 2003, 2004, Ward 2005).

Whatever its rationale, it is clear that the government is committed to building up a repertoire of 'relevant' research that it can draw on, and large scale policy evaluations have become one of its key tools in doing so. Policy initiatives pertaining to disability and disabled people's lives – from welfare-to-work programmes to housing – have been particularly affected by this trend toward evaluation. Some of the most high profile, and costly, evaluations since New Labour came to power in 1997 have related to the assessment of labour market initiatives aimed at disabled people, including the New Deal for Disabled People (NDDP) (Walker 2000, 2001). However, the potential role of geography to inform commissioned evaluations of such key disability policy initiatives has arguably been limited, with notable disconnections in terms of the disability policy areas that geography is perceived as relevant to, as well as a lack of geographical analysis in policy evaluations affecting disabled people (Imrie and Edwards 2007).

This chapter draws on UK government-commissioned evaluations of welfare-to-work initiatives for disabled people in the UK (and more specifically, NDDP) to examine some of these disconnections, and seeks to make a case for a 'spatially aware social policy' in terms of evaluations affecting disabled people's lives (Pinch 1998: 564, see also Mohan 2003). In so doing, it identifies some of the limitations of current evaluations in terms of the way space and spatial processes are constituted, and with what effect for their understanding of disability and the outcomes such programmes seek for disabled people. For some time now, geographies of disability have highlighted the way in which socio-spatial processes operating at different scales mediate, and shape, disabled people's lives and identities (Butler and Parr 1999, Gleeson 1999, Imrie and Edwards 2007). Meanwhile, geographers concerned with welfare reform have pointed to the significance of local labour markets and inter-scalar relations in explaining the diverse outcomes of welfare initiatives as they bed down in different localities (Haughton et al. 2000, Peck 1999, 2002, Peck and Theodore 2001, Sunley, Martin and Nativel 2001). However, these forms of analysis are frequently lacking in evaluations which view space as an irrelevance in terms of contextualising, and influencing, policy outcomes.

The chapter is divided into four main sections. I first discuss the rise of evidence-based policy and evaluation under New Labour, and go on to set this in the context of the UK welfare-to-work agenda for disabled people in the second section. The third section opens up a critique which debates the potential for analyses which take cognisance of the ways in which geography and geographical processes may contribute to an understanding of how workfare programmes operate, and drawing on government-commissioned research reports, focuses on this specifically in the context of the NDDP evaluation. I conclude by highlighting some of the consequences of the lack of spatial analysis for the way in which disabled people, and disability, are understood in such programmes.

Enter the Technocratic State: The Politics of Evidence-Based Policy and Evaluation

The use of social research to inform and evaluate policy programmes is not a new phenomenon, although historically different government administrations have demonstrated very different attitudes towards its potential role. Thus whilst social scientists were famously castigated, and social science research funding cut, in the Thatcher era, the inception of the Blair government in 1997 appeared to mark a sea change in the relationship between (social) research and government (Davies, Nutley and Smith 2000, Sanderson 2003, Walker 2001). In his oft-quoted speech to the Economic and Social Research Council, in 2000 for example, the then Secretary of State for Education, David Blunkett, stated the need for 'research which leads to a coherent picture of how society works' (Department for Education and Employment 2000: 22), stressing that 'social science research is central to the development and evaluation of policy' (24). Labour's approach to evidence-based policy has been described by some as 'anti-ideological' (Solesbury 2001: 6), insofar as it appears to be more concerned with pragmatism than politics.

Recognising that research is never conducted in a value-neutral vacuum, however, many commentators have criticised the evidence-based agenda and the way in which it has been employed by the government. For some, the notion of 'what works' ascribes research a utilitarian role, based on an instrumental rationality in which decision-makers (in this case, policymakers) are 'presented as seeking to optimise or utility-maximise in relation to clear, given goals' (Sanderson 2004: 368). In this sense, the relationship between research and policy is seen as a linear, and relatively uncontested one, based on the modernist assumption that (scientific) knowledge can be employed in a rational way to manage social and economic problems (Sanderson 2003, 2004). Such a notion is clearly evident in Blunkett's speech, when he states that 'rational thought is impossible without good evidence' (Department for Education and Employment 2000: 24).

Research in this context, then, is about practical application to a defined issue, rather than an intellectual or theoretical endeavour in its own right, and it is this notion which has led to conflicting opinions within the academe about how far social science should be servicing the needs of the government's evidence-based agenda. Thus whilst some geographers have called for the discipline to respond to the 'policy turn' by making geographical research more useful and policy-friendly (Martin 2001, Dorling and Shaw 2002), others have cautioned against such a move for its potential to lead to the 'belittling of particular epistemological positions, modes of enquiry, and forms of expression and writing' (Imrie 2004: 698). In particular, there is a concern that evidence-based policy has the potential to premise certain (and often, positivist) forms of knowledge over other forms of experience and modes of enquiry, and to narrow the frame of what gets researched or evaluated in the first place.

As a particular form of research, evaluation is necessarily caught up in the nexus of these debates, not least because it is, in and of itself, an applied form of enquiry concerned with assessing the outcomes and processes of a particular policy and/or initiative against some kind of yardstick, with a view to promoting its improvement (Weiss 1998). As Weiss (1998: 15) notes, what makes evaluation different from other forms of research is its concern with utility, with addressing questions which emanate from policy actors and/or communities, and its 'judgemental quality', which 'tends to compare "what is" with "what should be".' As in any academic community, there are divergent views amongst evaluators about how far evaluation can help to understand the ways in which policies impact on different groups and appropriate methodologies to apply in doing so. Many evaluators have been critical of the traditional experimental design, for example – the randomised control trial – for ignoring the different contexts in which policies and initiatives work, and for addressing 'too single-mindedly the question of whether a program works at the expense of knowing why it works' (Pawson and Tilley 1997: xv), and yet this is still aspired to by many in policy circles who see it as the gold standard of evaluation methodologies. It is within the context of these debates regarding the legitimacy of different types of knowledge, that the evaluation of Labour's welfare-to-work initiatives for disabled people is situated.

Disability and the Evaluation of the Welfare-to-Work Agenda

The welfare-to-work agenda has become one of the hallmarks of New Labour's tenure in government. Grounded in Blair's third way vision in which 'no rights without responsibilities' became the slogan of the day, the government has sought to promote active welfare, whereby rights to claim various out-of-work entitlements come with increasingly stringent requirements to actively seek work. In drawing distinctions between 'older' welfarism, and this approach, Peck and Theodore (2001: 429) suggest that 'workfare regimes are inherently more dynamic: they privilege transitions from welfare-to-work, typically through the combined use of "carrots" in the form of work activities, and job-search programmes and "sticks" in the form of benefit cuts for the non-compliant.' In attempts to tackle welfare dependency, addressing the employability of the inactive has therefore become a key goal, with participation in paid employment being constructed not just as a way out of dependency but also as a gateway to social inclusion (Department for Work and Pensions 2006, Hall 2005, Levitas 2005, Wilton and Schuer 2006).

Disabled people have been particularly affected by the shift towards policies and initiatives aimed at labour market activation. Shortly after its election in 1997, the government announced a number of New Deal programmes aimed at young people, lone parents, and disabled people. The broad aim of the programmes is to support those people who are unemployed or economically inactive and claiming benefits to find and sustain employment. Whilst each scheme is distinctive, the programmes having generally involved support from a personal advisor to assist with job search, benefits advice, and in-work support. In the case of disabled people, NDDP is aimed at people on incapacity-related benefits, and participation in the programme is voluntary. The programme was piloted through 12 NDDP Personal Advisor Pilots (six launched in 1998, six in 1999), and 24 Innovative Schemes designed to 'evolve and test new ways of helping disabled people to secure or remain in work' (Walker 2000: 314). This was followed, in 2001, by the national roll-out of Job Brokers across the country, organisations from the voluntary, private and public sector contracted by the government to deliver NDDP.

The NDDP, and related workfare initiatives, have been an opportunity for the Labour government to demonstrate its commitment to evidence-based policy. NDDP, for example was piloted and evaluated prior to national implementation, and has continued to be evaluated as Job Brokers were rolled out nationally. The evaluations could best be described as quasi-experimental in design; the research consortia responsible for the evaluation of the national extension of NDDP were proponents of using randomised controlled trials as part of the evaluation design, but this was rejected by ministers at the last minute on ethical grounds (although it has since been used in other workfare initiatives affecting disabled people). The evaluations seek to combine elements of both process and summative evaluation (i.e., the processes through which the programme is working, and the outputs/ outcomes respectively), and are large, multi-method affairs, which have spawned a substantial industry, if not a modern production line, of evaluation reports, commissioned by the Department for Work and Pensions (for a full list of the different components and evaluation reports, see Stafford et al. 2007).

The evaluations themselves have been undertaken by a large research consortium comprised predominantly of social policy researchers from universities and independent research institutes in the UK, but also economists based in the US who have been responsible for the cost-benefit analyses and impact assessments, indicating the desire of the government to learn lessons from American-style workfare programmes (Peck and Theodore 2001, Ferguson 2002). Reflecting notions that appeal to objectivity and value-neutrality in producing evidence-based policy, the consortium itself is recognised by the government as a respected and 'independent' source of evaluative data, and has been awarded successive contracts for evaluations of the different stages of NDDP. It is interesting in this context to note Oliker's (1994) study of the policy influence of a particular firm of evaluation researchers on the workfare agenda in the United States, in which she suggests that evaluations can legitimise certain policy ideas (in the case of her study, that women should work) which then become embedded as the basis for further evaluation, and narrow the types of questions that get asked about the impact of welfare reform agendas.

This has an applicability in the context of NDDP, where many disabled people themselves have criticised the workfare agenda for the way it demeans alternative contributions that can be made to society other than via the disabled person as 'worker-citizen', and for focusing too singularly on the supply side of the employment equation (Barnes 2000, Roulstone 2000, Barnes and Mercer 2005, Hall 2005). Indeed, in the context of the latter, it is interesting that little is said in NDDP about the type or quality of job that disabled people can expect to participate in through the programme. Thus, whilst promotional literature stresses the importance of matching individual skills to the right job, findings from the evaluations note that the largest proportion of all NDDP participants entering work, a quarter, found themselves in what was termed 'elementary occupations (or routine, unskilled occupations)', with the next largest group in sales and customer service roles (Stafford et al. 2007: 129). Concerns about job type and quality, however, appear to be less significant for the government, where some form of paid work is seen as better than none, and point to broader policy questions which can often be circumscribed in workfare evaluations that are based around the 'lifeless simplicity of the pure social experiment' (Oliker 1994: 209). It is with the aim of considering some of these broader policy questions that I go on to explore some of the missing (geographical) analyses in evaluations of disabled people's participation in NDDP.

Reclaiming the Absent Geographies of Disability in Workfare Evaluation

Geographers and geographical analysis have made a significant contribution to the understanding of the processes and consequences of welfare-to-work initiatives (see for example Peck 1999, 2002, Peck and Theodore 2001, Sunley, Martin and Nativel 2001, Smith et al. 2008), and within the geographies of disability, disabled people's experiences of the labour market and employment (Dyck 1999, Hall 1999, 2005, Wilton 2004, Wilton and Schuer 2006). Wilton and Schuer (2006: 186), for example, have sought to situate disabled people as workers within the context of neoliberal welfare reforms, highlighting the way they occupy a 'precarious position between an increasingly hostile welfare state and a labour market in which the "able-body/mind" remains a largely unquestioned norm' (see also Wilton 2004). Such studies point to the inter-scalar processes which define how disabled people themselves are understood as workers, and how employers understand what it means to create accommodating workspaces. They also illustrate how disabled people experience the workplace as a particular type of space or 'socially-valued site' (Wilton and Schuer 2006: 187) and the interrelationship between the workspace and the disabled body (see also Hall 1999).

Such analyses are key because they focus our attention on the nature of disabling spaces and broader spatial processes (such as the increased flexibility of labour markets) which impact on disabled people's lives, rather than on the disabled individual and their employability which, as discussed earlier, has become the focus of many welfare-to-work initiatives. Indeed, as disabled people and the disabled people's organisations have sought to argue, disability needs to be understood as a product of a disabling society and environment which presents barriers to disabled people, rather than as a consequence solely of an individual bodily impairment (Oliver 1990, Barnes and Mercer 2005).

Such forms of geographical analysis would however appear to be missing from official evaluations of labour market initiatives affecting disabled people, including NDDP, from which geographers have been noticeably absent. This may reflect the fact that some geographers do not wish to see themselves as servicing the needs of the state policy machine, but rather undertake what Pain (2006: 251) refers to as 'counter-policy research', critiquing policy from outside the state arena. However, as Imrie and Edwards (2007) argue, it may also reflect perceptions on the part of policymakers about the relevance of geography to the evaluation of policies affecting disabled people's lives, or the way in which certain welfare-to-work client groups become the domain of certain disciplines.

In the case of initiatives such as NDDP it is social policy analysts and economists (rather than geographers) who have become the key protagonists in welfare-to-work policy, reflecting what Pinch (1998: 556) suggests is the dominance of particular *knowledge communities* in policy realms. Defining such communities as 'the networks of power relations that enable certain communities of people to impose their ideas on others', he argues that up until recently, geography had relatively little impact on social policy, with geographers paying more attention to economic and regional policy.

Such issues are significant because, as Imrie and Edwards (2007: 624) argue in the context of disability, the limited impact of geographies of disability in public policy circles means that disciplines such as 'medicine, health studies, and social policy… collectively, drive forward both academic and popular conceptions of, and narratives about, disability'. Disabled people have long sought to challenge notions of disability which have emanated from medical and social policy professionals. In this chapter, I do not seek to assert geography as another dominant knowledge community, nor prioritise a spatial perspective above others. Rather, my concern is that government-commissioned evaluations of workfare initiatives aimed at disabled people, operating as they do on a technocratic, linear model, frequently de-contextualise space and spatial processes and their impacts for disabled people, with the result that the potential contribution of geographical analysis is frequently overlooked or misunderstood.

In the context of the welfare state, for example, Powell and Boyne (2001) suggest that geographical analysis has little to say about the outcomes of the welfare state, and that analyses to date of spatial outcomes have been problematic. Contesting their arguments, however, Mohan's (2003) assertion of the role of geography in understanding the welfare state and social policies is instructive. He draws attention to the need to focus on the uneven impact of national and international processes (such as globalisation and changes in the organisation of production and work practices) on the welfare state and on how social policies are addressing social needs/spatial inequalities. Crucially, he also points to the importance of geography in terms of its focus on *context* in exploring 'the difference that space makes to the operation of welfare systems' (Mohan 2003: 370). Studies in this area have drawn on different forms of analysis. On the one hand, geographers have mapped and analysed spatial disparities and inequalities inherent in the way policy initiatives play out across different regions. Sunley, Martin and Nativel's (2001) work on the New Deal for Young People (NDYP), for example, employs quantitative data to demonstrate the uneven impacts of NDYP across the UK. As they argue, this regional patterning is strongly related to the nature of local and regional labour markets, yet this is something which the government has been keen to deny in talking up the success of its welfare-to-work programmes.

Geographers have also drawn on qualitative, depth case study approaches to explore how processes happening in localities shape policy outcomes. Imrie's (1999) work on the politics of access for disabled people, for example, demonstrates how the provision of access to the built environment is mediated through the diverse institutional structures and political environments that characterise different local authority areas, and the existence of, and support for, local access groups (see also Imrie and Hall 2001). Similarly, Milligan's (2000: 53) work on geographies of caring demonstrates how macro-level changes in the nature of care policy (for example, the move away from residential care to community care) have created 'differential experiences of carers across space', depending on the localities within which they live. Thus, diversity in local demography, resource availability and the existence of formal/informal networks of support in different areas all mediate the experience of community care.

What characterises these studies is not just a focus on the locality to the exclusion of other scales of analysis. Rather, the locality is employed as a tool to explore how processes happening at different spatial scales (for example, national policy guidelines) get played out, and transformed, at the local level, and with what consequences for recipients of policy. In so doing, they often serve to highlight tensions which exist between national policy guidelines, and local implementation, something which is significant in the context of NDDP, where New Labour has placed a key focus on local flexibility in delivering welfare-to-work. More significantly, such studies work with a notion of space as a productive entity, whereby institutional and political processes transform, and are in turn transformed by, diverse spaces. It is in the context of these notions of productive space that I go on to consider the evaluations of NDDP.

'Best-fit Geographies': Abstracting Space in the Evaluation of NDDP

Geographers working in the area of welfare-to-work have highlighted tensions inherent in New Labour's attitude towards place as an explanatory variable in the success or otherwise of the New Deal programmes (Peck 1999, Sunley, Martin and Nativel 2001). In the context of NDYP, for example, Sunley, Martin and Nativel (2001: 485) suggest that there has been a failure by the government to recognise any spatial variability in terms of impact and that 'comments about spatial differences in the outcomes of the New Deal welfare-to-work programme have been firmly "off-message".' This lack of spatial analysis may be something that the government has sought to avoid (particularly because it places a focus on the uneven geography of local labour markets, which some have argued has undermined the success of the New Deal: see for example, Peck 1999). However, I would also argue that there is a limited conceptualisation of geography within the evaluations of such programmes, which views space as a backdrop to other processes, rather than an integral context with the potential to influence the way in which such programmes work, and their outcomes for disabled people.

An exploration of the design of the NDDP evaluations begins to highlight some of these issues. The evaluations of NDDP have been a two stage process. The first evaluation was concerned to assess the outcome of the pilot programme, including the Personal Advisor Service (PAS), whilst the second explored the national roll-out of Job Brokers across the UK. Twelve areas were selected as the basis for piloting and evaluating the PAS, with the evaluation itself having a wide-ranging number of objectives including exploring how the PAS operates and the type of assistance it offers, through to identifying outcomes for clients, including an assessment of the numbers helped into work and those who have 'improved their employability' (Walker 2000: 318).

The rationale behind piloting in particular geographical areas itself raises a number of questions about how space is understood in such programmes. In NDDP, the pilot areas for PAS were selected on the basis of a number of features: the type of area and labour market (rural, urban, mixed, inner city), and rates of unemployment/incapacity benefit claimants (low, medium, high) (Green, Owen and Hasluck 2001, Lessof et al. 2001, Loumidis et al. 2001). One of the methods the evaluators claimed to use in the study – amongst large scale surveys of participants and non-participants on the programme, employers, and qualitative work with Personal Advisors themselves and other stakeholders – were 'area studies'. These studies consisted of a brief pen picture in the evaluation report of each area and an analysis, based on administrative data, of the labour market characteristics of the area to 'provide a description and limited assessment of the context in which the New Deal for Disabled People operated' (Loumidis et al. 2001: 251). Thus, Sandwell, one of pilot areas deemed to have higher than average levels of unemployment, is described as 'a heavily urbanised area in the West Midlands conurbation', in which 'the industrial base rests heavily on manufacturing...Associated with this is a marked concentration of employment in manual occupations, while professional and managerial occupations are under-represented relative to the national average' (Loumidis et al. 2001: 260). Meanwhile, South Devon – a 'medium unemployment/inactivity area' – is described as being 'a typical resort and retirement area with an older than average population profile...A greater than average share of employment in personal and protective service occupations underlines the importance of tourism in the local economy' (Loumidis et al. 2001: 263). What is significant however, is that beyond an appendix describing and comparing the areas at the back of the report, nowhere do these, and the economic, social, or political processes that constitute them, come into play in the analysis of, for example, disabled people's experiences of the service, or in the effectiveness of personal advisors' relationships with employers. Rather, the findings are abstracted out of their spatial context, with the selection of areas being employed for little more than sampling purposes.

The use of geographical area in this way is also evident in the means by which such evaluations seek to assess the net impact of the programme, otherwise known as establishing the counterfactual (the situation that would have occurred in the absence of the programme) (Walker 2000). This involves establishing a control group against which to compare outcomes with those who received the intervention. Using what is termed 'matched areas' has become a firm favourite in evaluations of welfare-to-work initiatives for the Labour government, whereby control areas with similar characteristics to the pilot areas are selected as the basis for comparison (Walker 2001). In NDDP, once random assignment was ruled out, this comparison was undertaken by commissioning a national survey of incapacity benefit claimants not participating in NDDP, who were sampled on the basis of living in similar types of area, and having similar characteristics to, those in the NDDP pilot areas.

Such examples are indicative of a rather one-dimensional view of space, in which areas are seen as little more than holding vessels for people with particular characteristics or a canvas across which different variables are to be mapped. They fail to recognise the dynamic qualities of space and spatial processes, or how areas are intimately interconnected at a range of spatial scales; the notion of selecting pilot and control areas would suggest that spaces are independent, stand-alone entities which can be set up against one another as a basis for experimentation. In promoting his notion of a spatially aware social policy, Pinch (1998: 565) states that 'It is important to stress that...space is not conceptualized as an empty "container" through which social processes operate but an important constituent of those processes'. It is however, precisely this 'container' approach that would appear to characterise the design of the NDDP pilot evaluation.

Indeed, within the discussion of the analysis of the labour market characteristics of the areas selected, there is an obsession with trying to obtain administrative data that matched geographical boundaries. For example, the pilot areas were based on Benefit Agency administrative districts, which were not necessarily coterminous with local labour market areas and the evaluators struggled to compile data on the necessary variables to create an accurate picture of the local labour market; they therefore had to be satisfied with what they referred to as '"best-fit" geographies' in terms of the data that could be brought together (Lessof et al. 2001: 111). It is surprising, then, that despite this analysis, there is so little discussion of the impact of local/regional labour markets on the findings, or disabled people's experience of finding work. For whilst supply-side issues have become the focus of current policy (for example, in addressing barriers that disabled people face in accessing the labour market due to health problems, or motivational issues), it is well-recognised that the state of local/regional labour markets and labour market demand (including types of jobs available) are the other side of the equation. Thus, analysis has shown that disabled people's employment rates are often strongly associated with the nature of local labour markets, and that in areas with depressed labour markets, disabled people are likely to find themselves 'further down the "labour queue"' (Meager and Hill 2005: 29, see also Howard 2003, Beatty, Fothergill and Macmillan 2000). Ironically, qualitative research with employers as part of the evaluation of the national roll-out of NDDP highlighted this point, with some small employers stating that they were 'less likely to recruit someone with a disability or health condition if they could hire fairly easily locally' (Stafford et al. 2007: 125).

Despite presenting different labour market areas as a backdrop to the PAS pilot evaluation, however, the evaluation reports do not specifically examine connections between the labour market characteristics of the different areas and disabled people's experiences, and therefore fail to take seriously the context of localities in attempting to 'learn lessons' from the programme. This is perhaps ironic given that New Labour's workfare agenda has been based around 'decentralised programming', in which local experimentation is the order of the day, rather than top-down control via national policy/agencies (Peck and Theodore 2001: 446; also Haughton et al. 2000). The piloting of PAS in different areas would have seemed to present an opportunity for an assessment of spatial variation at the local level in how NDDP operated, a setting in context of the programme around features such as the local labour market, the local political and institutional contexts which framed relationships between different stakeholders involved in NDDP, and even the local politics of disability (particularly as some disability organisations were themselves contracted to act as Personal Advisors and subsequently, Job Brokers). Statements in the evaluation reports of the national Job Broker programme allude to some of these points, albeit in a way that is de-contextualised in terms of the locality being referred to; for example, they note that amongst the Job Brokers, 'there were examples of organisations with relatively sparse local contacts…particularly if they were new to disability and/or employment' (Corden et al. 2003: 14), whilst others had a well-established history of working in the disability arena in particular areas.

Yet as Peck and Theodore (2001) suggest, the concern with implementing new ideas at breakneck speed for political reasons has often overridden any serious learning from pilots, and hence localities. Walker (2001: 322) also notes of the New Deal evaluations, that there was no scope for what he terms 'comparative ethnography' which would have shed some light on the differences within, and between areas, as it was deemed too costly. The abstraction of findings from any sense of spatial context in NDDP would therefore seem to be in part a product of an evaluation driven by offering an overall message about what works – a technocratic fix 'focusing on the decontextualised, disembedded methodological essence of workfare programmes' (Peck and Theodore 2001: 432) – rather than an assessment of local differentiation.

An assessment of how geography is employed in the findings of the NDDP evaluations as an explanatory 'variable' is also instructive in this context. Interestingly, many of the reports that make up both the evaluation of PAS, and the national study of Job Brokers hint at the potential difference that spatial processes can make to the relationship between disabled people, Job Brokers, and employers in NDDP – whether this be the finding that geography and location of Job Brokers' services was a factor affecting the likelihood of NDDP participants obtaining a job, or participants' views that being able to work in certain environments, particularly from home, would act as a 'bridge' to get them back into the labour market (Stafford et al. 2007). However, there is little conceptualisation of these processes or recognition of how disabled people's lives are intimately bound up with, and shaped by, social-spatial environments. One specific example of this is the finding to come out of the overall national evaluation of NDDP (based on the Job Broker service) that region was a key factor in affecting how likely a disabled person was to move into paid work (Stafford et al. 2007). This finding was based on multivariate analysis of survey data, but as the report goes on to say, 'other than showing that there was a significant regional variation, the findings are inconsistent across cohorts and difficult to interpret, as there is no obvious association with regional labour markets' (Stafford et al. 2007: 126; see also Kazimirski et al. 2005). Thus the relationship between geography and disabled people's movements into work is deemed to be complex and the evidence 'mixed' (Stafford et al. 2007: 125). Such statements are partly a product of quantitative analysis, whereby location is seen as just another variable to be thrown into the statistical pot, and are indicative of an analysis which is concerned with presenting an overall message about particular outcomes from the programme which can be picked up and generalised at the national level.

Elements of the qualitative analysis nevertheless hint at some of the lived geographies of disability in terms of disabled people's relationship with the Job Broker service, and their feelings about accessing the service. Thus, many Job Brokers in rural areas spoke of the difficulty for disabled people in terms of travelling to access their services; issues were also raised about the location of Job Brokers' offices and the importance of not locating within, or next to, JobCentres, in attempts at 'avoiding some of the stigmatisation thought to attach to the jobcentre' (Corden et al. 2003: 15). Such examples illustrate what Dyck and O'Brien (2003) refer to as the personal geographies of disabled people, and how they negotiate moving in and through different spaces, whether it be travelling to a Job Broker, or – in reflecting the way in which spaces can be inscribed with particular values – entering a space which may confer an identity which is demeaning.

These personal geographies are also evident in an example from one of the reports of the experience of a disabled woman at a particular site, the Job Broker's office. The woman had registered with a Job Broker, but subsequently de-registered, because:

> she felt uncomfortable in their offices. She complained that they were located at the top of a building and she had had to wait for the lift to be unlocked, that she had been kept waiting before anyone attended to her, that the chairs were uncomfortable and then when she was eventually seen, she was given a form to complete, but no-one asked if she needed any help in completing it (Pires et al. 2006: 97).

This example demonstrates how the woman's (bodily) encounter with the Job Broker had been particularly affected by the spatiality of the office environment. It points to the micro-geographies which mediate disabled people's experience of everyday life, and the multi-scalar ways in which place plays a role in impacting on disabled people's relationship with employment, from the contexts of local labour market through to everyday workplace environments, all things which arguably require greater attention in evaluations of programmes such as NDDP.

Conclusions

This chapter has sought to encourage discussion regarding the potential role of geography in shaping evaluations of workfare initiatives that impact on disabled people's lives. I suggest that evaluation studies such as that of the NDDP programme necessarily abstract the presentation of findings from any spatial (and in particular, local) context, reducing location to either a statistical variable, or the basis for a sampling strategy. The missing geography of the NDDP evaluation – or of any sense of spatial context – is potentially a consequence of the vast number of reports which have been produced on the different elements of the evaluation programme, making it difficult to connect findings into a coherent, and contextualised, whole. However, it is also a consequence of an under-conceptualised notion of geography, which fails to acknowledge the importance of multi-scalar spatialities as a context to such programmes and disabled people's relationship with the labour market.

That this is the case is reflective of a concern by the government and policymakers to focus on implementation and glean ideas not just about 'what works', but about what 'works better' in the evidence-based agenda. To provide fully locally-contextualised studies would be to invite greater complexity, and to challenge notions of transferability of findings; what the government wants is a clear message about what it needs to do. Peck (2002: 352) summarises this well when he suggests that: 'Dominant evaluation methods…which are based on quantitative, control-group studies, have played a role not only in generalising decontextualised and essentialised policy knowledges about workfare programs and their rationally-acting participants, but also in privileging assessment metrics oriented to short term outcomes and the minimisation of costs.' Assessing cost effectiveness is indeed a key aim of the NDDP evaluation agenda, with the cost-benefit analysis being calculated in the context of the government (money saved in terms of welfare payments, for example), NDDP registrants themselves and society (Greenberg and Davis 2007).

Arguably, one of the consequences of this focus on cost-effectiveness is the way in which disabled people and disability become constructed in terms of an economic cost to society. Even the language by which disabled people are referred to in the evaluation reports – as either the 'eligible population', NDDP 'registrants', or programme 'participants' – seems to abstract disabled people from any sense of their lived world/experiences; rather, their identities are recast in terms of technocratic programme-speak. Economists' assessments of costs-of-illness estimates are bound up in notions of lost productivity or increased spending on welfare benefits. There is less consideration of, for example, the cost to a person of discrimination or, as Wolff (2007) notes in the context of mental illness, the social construction of illness which underpins such estimates.

NDDP's focus on disabled people as latent economic potential moves us away from a discussion of the broader barriers and realities which inform disabled people's everyday lives. Thus, it is interesting to note that whilst the cost-benefit analysis recognised that income was only one facet of potential costs or benefits for NDDP registrants, the evaluators were able to say little about the overall well-being of disabled people as a result of participation in NDDP due to 'scant information' (Greenberg and Davis 2007: 5). Despite the large scale surveys of those eligible for NDDP, and NDDP registrants, one does not get a clear sense of disabled people's voices within the evaluation or how the programme may have impacted on people's lives.

The potential for a geographical incursion into workfare evaluation therefore seems to suggest itself in a number of ways. Firstly, geographers can explore regional disparities in outcomes of NDDP for disabled people, to provide a macro-scale picture of how NDDP is playing out across the UK. Perhaps more significantly, however, is the need for a locally embedded analysis, which draws on in-depth case study methodologies, and has the ability to connect multi-scalar processes operating in local areas – such as labour market conditions, local political environments, and the nature of diverse Job Brokers – to provide a more coherent and contextualised picture of how NDDP gets worked out on the ground, and with what impact for disabled people. Such an approach may arguably create an opportunity to, if not reclaim the voice of disabled people as participants in NDDP, at least redress some of the technocratic imbalance in such evaluations, and refocus the analysis on to structural and institutional barriers in society which shape disabled people's access to the labour market.

How geographers engage in this policy arena is another debate. Much ink has been spilt over the last few years in discussing how (if at all) geographers should engage with the government policy machine, but as Ward (2005) helpfully points out, engagement with policy does not necessarily mean engaging solely with government officials, but also activist groups operating within and outside formal policy networks, and the public. There are a number of disability groups, for example, that the government consults on these issues and there may be potential for alliances in so far as geographers seek to move the 'work agenda' away from a pure focus on the employability of disabled people. Yet even within the cosy world of government commissioned evaluation consortia, there are chinks of light, opportunities for engagement. In a recent paper published in the journal *Evaluation*, Robert Walker, one of the architects of the NDDP evaluation, offers up for debate a new way of approaching workfare evaluation, based on the principle of entropy. Whilst I do not wish to launch into an account of thermodynamics from which this principle comes, nor its rights and wrongs as a basis for evaluation, it is interesting that he critiques some of the features of current evaluations, such as the use of counterfactuals which he argues are artificial measures that can often tell us little. The metaphor of entropy, he suggests, 'emphasises the need for evaluations to be simultaneously conducted at different institutional levels and from different perspectives' and 'presumes that outcomes are contingent on history, location and circumstance' (Walker 2007: 194). In calling for a more contextualised, embedded, form of evaluation, perhaps the time for geographers to respond has come.

Acknowledgements

I would like to thank the editors of this volume for their valuable comments on an earlier version of this chapter.

References

Barnes, C. 2000. A working social model? Disability, work and disability politics in the 21st century. *Critical Social Policy*, 20(4), 441-57.

Barnes, C. and Mercer, G. 2005. Disability, work and welfare: challenging the social exclusion of disabled people. *Work, Employment and Society*, 19(3), 527-545.

Beatty, C., Fothergill, S. and Macmillan, R. 2000. A theory of employment, unemployment and sickness. *Regional Studies*, 34(7), 617-630.

Butler, R. and Parr, H. (eds) 1999. *Mind and Body Spaces: Geographies of Illness, Impairment and Disability.* London: Routledge.

Corden, A., Harries, T., Kellard, K., Lewis, J., Sainsbury, R. and Thornton, P. 2003. *New Deal for Disabled People National Extension: Findings from the First Wave of Qualitative Research with Clients, Job Brokers and Jobcentre Plus Staff.* Department for Work and Pensions (DWP) Research Report 169. Sheffield: DWP.

Davies, H.T.O., Nutley, S.M., Smith, P.C. (eds) 2000. *What Works? Evidence-Based Policy and Practice in Public Services.* Bristol: Policy Press.

Department for Education and Employment, 2000. *Influence or Irrelevance: Can Social Science Improve Government?* Secretary of State's ESRC Lecture Speech 2 February. London: DfEE.

Department for Work and Pensions, 2006. *A* New *Deal for Welfare: Empowering People to Work.* Green Paper. London: The Stationery Office.

Dorling, D. and Shaw, M. 2002. Geographies of the agenda: public policy, the discipline and its (re)'turns'. *Progress in Human Geography*, 26(5), 629-646.

Dyck, I. 1999. Body troubles: women, the workplace and negotiations of a disabled identity, in *Mind and Body Spaces: Geographies of Illness, Impairment and Disability*, edited by R. Butler and H. Parr. London: Routledge, 119-137.

Dyck, I. and O'Brien, P. 2003. Thinking about environment: incorporating geographies of disability into rehabilitation science. *The Canadian Geographer*, 47(4), 400-413.

Ferguson, R. 2002. Rethinking youth transitions: policy transfer and new exclusions in New Labour's New Deal. *Policy Studies*, 23(3/4), 173-190.

Gleeson, B. 1999. *Geographies of Disability.* London: Routledge.

Green, A., Owen, D. and Hasluck, C. 2001. *New Deal for Disabled People: Labour Market Studies*. Department for Work and Pensions In-house Report No. 79, Social Research Branch. London: DWP.

Greenberg, D. and Davis, A. 2007. *Evaluation of the New Deal for Disabled People: The Cost and Cost-Benefit Analysis*. Department for Work and Pensions Research Report No. 431. Leeds: Corporate Document Services.

Hall, E. 1999. Workspaces: refiguring the disability-employment debate, in *Mind and Body Spaces: Geographies of Illness, Impairment and Disability,* edited by R. Butler and H. Parr. London: Routledge, 138-154.

Hall, E. 2005. The entangled geographies of social exclusion/inclusion for people with learning disabilities. *Health and Place*, 11(2), 107-115.

Haughton, G., Jones, M., Peck, J., Tickell, A. and While, A. 2000. Labour market policy as flexible welfare: prototype employment zones and the new workfarism. *Regional Studies*, 34(7), 669-680.

Imrie, R. 1999. The role of access groups in facilitating accessible environments for disabled people. *Disability and Society*, 14(4), 463-82.

Imrie, R. 2004. Urban geography, relevance, and resistance to the 'policy turn'. *Urban Geography*, 25(8), 697-708.

Imrie, R. and Edwards, C. 2007. The geographies of disability: reflections on a sub-discipline. *Compass Geography* [Online], 1(3), 623-640. Available at: http://www.blackwell-compass.com/subject/geography/section_home?volume=2&Go=Go&section=geco-social-geography [accessed: 22 August 2008].

Imrie, R. and Hall, P. 2001. *Inclusive Design: Designing and Developing Accessible Environments*. London: Spon Press.

Kazimirski, A., Adelman, L., Arch, J. et al. 2005. *New Deal for Disabled People Evaluation: Registrants Survey Merged Cohorts (Cohorts One and Two, Waves One and Two)*. Department for Work and Pensions Research Report No. 260. Leeds: Corporate Document Services.

Lessof, C., Becher, H., Corden, A. et al. 2001. *Evaluation of the New Deal for Disabled People Personal Adviser Service Pilot – Technical Report*. Department for Work and Pensions In-house Report, 89, Social Research Branch. London: DWP.

Levitas, R. 2005. *The Inclusive Society? Social Exclusion and New Labour.* Basingstoke: Palgrave Macmillan.

Loumidis, J., Stafford, B., Youngs, R. et al. 2001. *Evaluation Of The New Deal For Disabled People Personal Adviser Service Pilot*. Department of Social Security Research Report, No.144. Leeds: Corporate Document Services.

Martin, R. 2001. Geography and public policy: the case of the missing agenda. *Progress in Human Geography*, 25(2), 189-210.

Meager, N. and Hill, D. 2005. *The Labour Market Participation and Employment of Disabled People in the UK*. Brighton: Institute for Employment Studies.

Milligan, C. 2000. 'Bearing the burden': towards a restructured geography of caring. *Area*, 32(1), 49-58.

Mohan, J. 2003. Geography and social policy: spatial divisions of welfare. *Progress in Human Geography*, 27(3), 363-374.

Oliker, S. 1994 Does workfare work? Evaluation research and workfare policy. *Social Problems*, 41(2), 195-213.

Oliver, M. 1990. *The Politics of Disablement*. Basingstoke: Macmillan.

Pain, R. 2006. Social geography: seven deadly myths in policy research. *Progress in Human Geography*, 30(2), 250-259.

Pawson, R. and Tilley, N. 1997. *Realistic Evaluation*. London: Sage Publications.

Peck, J. 1999. New Labourers: making a New Deal for the 'workless class'. *Environment and Planning C*, 17(3), 345-372.

Peck, J. 2002. Political economies of scale: fast policy, interscalar relations, and neoliberal workfare. *Economic Geography*, 78(3), 331-360.

Peck, J. And Theodore, N. 2001. Exploring workfare/importing welfare-to-work: exploring the politics of Third Way policy transfer. *Political Geography*, 20(4), 427-460.

Pinch, S. 1998. Knowledge communities, spatial theory and social policy. *Social Policy and Administration*, 32(5), 556-571.

Pires, C., Kazimirski, A., Shaw, A., Sainsbury, R. and Meah, A. 2006. *New Deal for Disabled People Evaluation: Survey of Eligible Population, Wave Three.* Department for Work and Pensions Research Report No. 324. Leeds: Corporate Document Services.

Powell, M. and Boyne, G. 2001. The spatial strategy of equality and the spatial division of welfare. *Social Policy and Administration*, 35(2), 181-194.

Roulstone, A. 2000. Disability, dependency and the New Deal for Disabled People. *Disability and Society*, 15(3), 427-443.

Sanderson, I. 2003. Is it 'what works' that matters? Evaluation and evidence-based policy-making. *Research Papers in Education*, 18(4), 331-345.

Sanderson, I. 2004. Getting evidence into practice: perspectives on rationality. *Evaluation*, 10(3), 366-379.

Smith, F., Barker, J., Wainwright, E., Marandet, E. and Buckingham, S. 2008. A New Deal for lone parents? Training lone parents for work in West London. *Area*, 40(2), 237-244.

Solesbury, W. 2001. Evidence-based policy: whence it came and where it's going. *ESRC UK Centre for Evidence Based Policy and Practice: Working Paper 1.* London: ESRC Centre for Evidence Based Policy and Practice.

Stafford, B. with others. 2007. *New Deal for Disabled People: Third Synthesis Report – Key Findings from the Evaluation.* Department for Work and Pensions Research Report No. 430, Leeds: Corporate Document Services.

Sunley, P., Martin, R. and Nativel, C. 2001. Mapping the New Deal: local disparities in the performance of welfare-to-work. *Transactions of the Institute of British Geographers*, N.S. 26(4), 484-512.

Walker, R. 2000. Learning if policy will work: the case of the New Deal for Disabled People. *Policy Studies*, 21(4), 313-332.

Walker, R. 2001. Great expectations: can social science evaluate New Labour's policies? *Evaluation*, 7(3), 305-330.

Walker, R. 2007. Entropy and the evaluation of labour market interventions. *Evaluation*, 13(2), 193-219.

Ward, K. 2005. Geography and public policy: a recent history of 'policy relevance'. *Progress in Human Geography*, 29(3), 310-319.

Weiss, C.H. 1998. *Evaluation: Methods for Studying Programs and Policies.* New Jersey: Prentice Hall.

Wilton, R.D. 2004. From flexibility to accommodation? Disabled people and the reinvention of paid work. *Transactions of the Institute of British Geographers*, NS 29, 420-32.

Wilton, R.D. and Schuer, S. 2006. Towards socio-spatial inclusion? Disabled people, neoliberalism and the contemporary labour market. *Area*, 3(2), 186-195.

Wolff, N. 2007. The social construction of the cost of mental illness. *Evidence and Policy*, 3(1), 67-78.

Chapter 10

Placing Little People: Dwarfism and the Geographies of Everyday Life

Robert J. Kruse, II

Introduction

Dwarfism continues to be a source of fascination and confusion in contemporary society. From mysticism to freak shows, it is an identity of contradictions – an identity that has located adults of extremely short stature in a curious array of places including royal courts, circuses, and institutions. I maintain that the 'difference' of dwarfism is essentially a geographical one. First, it is a set of conditions that affect the physical negotiation of everyday spaces and places. Simultaneously, the ways in which dwarfism is understood in cultural terms relegate people of extremely short stature to particular spaces and places. Hence, both the physical and cultural aspects of dwarfism cause people to be 'placed' due to their body type.

Considered by many to constitute a physical disability, dwarfism can be situated within various traditions of geographical research on disabilities. In general terms, geographical approaches to physical disability fall into two categories (Park et al. 1998). One of these categories consists of research on the conditions of particular disabilities in quantitative terms of statistical analysis. This type of research has, for example, provided insight into the distribution of occurrences of specific disabilities and the provision of public services. Another body of geographical research has focused upon the socio-spatial experiences of people with disabilities (e.g., Chouinard 1999, 1997, Imrie 2001, 1998, Kruse 2007, 2003, 2002). These studies deal with issues of identity, space and social justice, and frequently use social theory to situate discussions.

This chapter addresses dwarfism within geographies of identity and social space. It draws upon interviews and participant-observations with a particular married couple in which the husband and wife are both little people. Although they reside in a rural area in the United Sates, a town in which they are the only little people, they remain linked via the Internet to a broader community of little people. Part of a larger research project on dwarfism, this chapter focuses on three geographical aspects of dwarfism. First it explores ways in which public space is experienced differently by people with dwarfism. Relatedly, it describes ways in which these little people arrange and adapt their private residential spaces in ways that challenge and reflect dominant norms of height. Finally, it follows the narratives of a particular family of little people who reveal the ways in which their identities and the corresponding spaces in which they are lived are mutually constitutive.

Theoretical Guideposts: Identity, Discourse and Space

Before proceeding any further, it is useful to define several key concepts and terms integral to this discussion. First, one of the most important issues for the participants in this research was the terms commonly used to label them. Rather than accepting the labels others have historically used to identify them, little people are increasingly using modern mass media technologies such as the Internet to represent themselves. Hence, the consensus among little people with whom I have spoken is that the preferred term of people commonly referred to as 'dwarves' is *little people*. The reason for this, I was told, is that the term is simply accurate and carries less entangled and derogatory cultural baggage than the word *dwarf*, or even worse, *midget*.

My use of the term *social space* draws from several theoretical perspectives of identity and space. Drawing from the ideas of post-structuralism, I view the changing placement of dwarfism in society as a result of shifting *discourses*. Here, several assumptions characteristic of post-structural thought as forwarded by Foucault, Barthes and others are helpful. From a post-structural perspective, discourses can be understood as groups of statements that structure the way we think about something, and the way we act on the basis of that thinking (Rose 2001). Because there is no essential definition of dwarfism, the historical placement of people with dwarfism has been the result of competing cultural discourses that have manifest in particular places and spatial practices. Furthermore, dominant discourses of dwarfism of any particular era have resulted from the authority with which, for example, medicine, academia or the entertainment industry have been vested. Hence, culturally dominant discourses of dwarfism have been derived from statements made by those individuals and institutions with the power to define dominant notions of normalcy, deviance, freakishness, comedy and mysticism. More than simply 'ideas' or 'concepts', dominant discourses have material implications in terms of the socio-spatial experiences of little people and their perceived 'place' in society. Here the notion of 'intertexuality', a term first coined by Kristeva in the 1960s and widely used in cultural studies, is especially important. The premise of intertextuality is that every cultural text (such as the statements that support particular discourses) is ascribed with meaning in relation to other cultural texts. For example, historically-specific cultural discourses of deviance are related to discourses of what was considered normal at those times. In other words, dwarfism has been defined and spatialized in relation to other dominant cultural discourses, many of which have little if anything to do with physical stature.

In some respects, the socio-spatial experiences of little people show similarities to other research on spatialized identities. For example, Valentine (1993) has observed the ways in which the '(hetero)sexing of space' reinforces the taken-for-granted way in which asymmetrical couples and families interact in public space. I maintain that a similar 'staturization of space' reinforces the dominant preference for able bodies of average height. Similar to the (hetero)sexing of space, the staturization of space is a result of the material environments (spaces that are obviously scaled for adults of average height) that clearly anticipate the presence of some individuals, but not others. Such environments, in addition to offering practical challenges to little people, also reinforce the othering of individuals whose bodies do not conform to the norm. Hence, both the material environment and the perception of little people are mutually constitutive. Moreover, the staturization of space involves issues of power in terms of social policy and those who are able to construct particular environments that are accessible or inaccessible to little people.

Methodological Considerations

As Pile (1991) notes, since the 1980s there has been considerable interest in the subjective experience of space. Underscoring the need to listen to the stories of less visible actors with disabilities, Butler (1994) specifically notes the need to consult with individual disabled persons in order to understand their spatial experiences. Situated within the interests of 'new cultural geography', this geographical examination of dwarfism required research techniques that enabled me to explore the realities of everyday life as they are experienced and explained by little people (Pile 1991). I maintain that such accounts offer a response to tendencies to objectify and misrepresent people with dwarfism. Consequently, this research is an exercise in interpretative geography and draws from traditions of feminist research. Such traditions include the importance of identifying both researchers and 'subjects' as active agents in the production of knowledge (Nagar 1997). It is at this point that reflexivity is especially important, and the understanding that the task of representing others is fraught with power relationships (Pile 1991). Hence, if this project is to be the result of 'disciplined subjectivity', then it is necessary to identify sources of bias (LeCompte 1987). I offer that I am a white, male academician of average height and have limited experience as being a member of any social minority. Not surprisingly, I was told by participants in this research that my status as an adult of average height significantly limited my understandings of the experience of little people. Yet, while my personal characteristics may effectively locate me on the social landscape of the non-disabled population, their disclosure also runs the risk of 'self-stereotyping' (Robertson 2002). Clearly, subjecting the participants to academic analysis run its risks – including depicting their accounts in ways that deliberately conform to particular theoretical perspectives. With these methodological issues in mind, I have made every effort to stay as close to the narratives of the participants as possible with the full understanding that I am providing only a partial perspective on the geographies of dwarfism.

My focus on a few individuals with dwarfism cannot in any way be construed as a representative sample. However, consistent with England (1993), my purpose is not to obtain empirical generalizations, rather it is to develop an in-depth understanding of the particular interviewees' frames of reference and socio-spatial meanings. Baxter and Eyles (1997) suggest that the credibility of qualitative research can be judged, in part, by the degree to which it provides an authentic representation of experience. Here, I note that the claims made about the socio-spatial experience of these participants came directly from the interviews. Baxter and Eyles (1997) also suggest that transferability is a means by which the relevance of research can be ascertained. Here, I note, that although dwarfism shows some unique socio-spatial characteristics specific to short stature, the accounts of the participants show issues of identity, disability and space consistent with other disability geographies. Also, Wilton and Cranford (2002) contend that, while a potential drawback of case studies is the inability to generalize, the relevance of such studies can be broadened by the application of particular theoretical frameworks. In this case, as mentioned earlier, I draw from post-structuralism and the ways in which competing discourses of identity produce different spaces. By doing so, I propose that this work can also be related to research on the discursive spatialities of other identities, including gender, race, and sexuality.

Meeting Paul and Mary Jamison

Through my discussions with little people early in this research, I obtained the email address of the president of a regional division of Little People of America, a national support and advocacy group for people with dwarfism. His initial email response was courteous, although he had questions about how I would use any information he provided. Too often, he told me, people of average height have attempted to gain access to little people for purposes of exploitation. However, in a subsequent phone call during which I described the exploratory purpose of my project and my academic affiliations, he agreed to be interviewed with his wife. Shortly thereafter, we arranged for a face-to-face interview at a local restaurant near their home.

On the day of the interview, I waited in the restaurant and watched the arrivals and departures in the parking lot. Finally, a large four-door sedan pulled in. The driver's side door opened slowly and a small man in his mid-forties slid sideways, perched on the edge of the seat and jumped to the ground. He closed the door and made his way around to the passenger's door and opened it with both hands. A small woman, about his age, gingerly extended her short legs to the ground and, after a little jump, stood aside as he closed the door, again using both of his hands. The way he put his weight against the car door as he closed it indicated that it was heavy for him. Then they walked side by side toward the building. The man opened the front door, pulling with both arms, and followed the woman inside. The hostess greeted them by name and with a smile of familiarity.

As I stood and introduced myself, I became keenly aware that the primary issue was body difference in both physical and social terms. For example, as we made initial small talk, I was very conscious of their looking up at me which made me uncomfortable. While standing up for an introduction is a common display of respect, in this case, it seemed to interfere with our exchange. Yet, after we were seated, I noticed that the difference in height was de-emphasized. In this position, we were able to view each other eye to eye which seemed to put both the Jamisons and me at ease for conversation. After about 30 minutes of chat, the conversation shifted to dwarfism.

Being Little in Public Space

Paul was 42 years old at the time of our first interview. Standing four feet six inches tall, he was employed as a computer programmer. Mary was 34 years old, four feet one inch tall, and worked as an administrative assistant at a local college. Paul explained that his condition, achondroplasia, is among the most common types of dwarfism. His limbs are shorter than average and his head appears a bit large in proportion to the rest of his body. Mary's body proportions are the result of Kniest Syndrome dwarfism. In contrast to Paul's body, Mary has a short torso with average-size limbs. She began by telling me that dwarfism is not a single physical condition, but refers to a group of genetic and hormonal conditions that result in extremely short stature in adults. She noted, however, that people of average height often perceive this variation monolithically through discourses rooted in folklore:

> A lot of people have never met a little person and they form their opinions on limited information. They've seen them in story books they've read, but they might not know that there really are little people out there. They think its fiction.

Paul added, 'Often we're portrayed as evil or strange, rather than professional and normal.' In addition to being affected by these discourses, Paul and Mary noted that there are relatively few opportunities to challenge the related notion that little people are 'abnormal' and out of place in ordinary situations of everyday life. Hence, the relatively small number of little people (approximately one in 10,000 births) increases the element of surprise for average-height people in public space and contributes to little people being perceived as different or 'others.' Due to such infrequent interaction between average-height people and little people, stereotypical representations, positive or negative, often go unchallenged (Sibley 1995).

Because of their atypical bodies, Paul and Mary explained how they do not blend in with the public in public space. The relative anonymity and invisibility available to average height adults on, for example, a bustling city street is unavailable to them. The Jamisons are always conspicuous – they can never pass for being of average height. Paradoxically, this hyper visibility contributes to feelings of alienation – especially on days when they encounter insensitive and intrusive comments and stares in public. The Jamisons noted that in many situations people of average height are courteous, although they may stare. However, Paul mentioned the teasing comments of teenagers he sometimes encounters at the shopping mall. He said his response depends upon his mood. Sometimes he ignores the comments, other times he is more confrontational adding something like, 'What's your problem?' Either way, he is not surprised by such encounters. He knows that there is always the possibility of harassment when he is in public spaces. Generally, both Paul and Mary are forgiving of the unwanted attention they receive.

In addition to the unwelcome comments and stares directed at Paul and Mary, their self-consciousness is increased in situations where they are perceived as less-able or even child-like. For them, negotiating simple everyday tasks such as using an automated teller machine (ATM), getting on a bus, reaching for an item on a store shelf, or opening a heavy door can be especially challenging. In such situations, it appears that discourses of dwarfism and disability are compounded by their intersections with gender and age identities. Because these situations are designed for average-height people, an otherwise capable and mature little person, especially a woman, may be perceived as childlike in relation to the physical environment. The infantilizing treatment that Mary describes seems to be commonly experienced by adult little people (Gerber 1996). It suggests an ambiguous identity in which a little person, like a 'cute child', is culturally located as an object. Furthermore, the little person, if viewed as 'cute', unintentionally transgresses spaces intended for children and adults (Merish 1995).

Relatedly, Paul and Mary also experience annoying violations of their personal space in public. Mary explained:

> A lot of times I'll be at the store, out shopping by myself, and I'll be at the register. Then someone behind me will speak over me and try to be next in line. They'll just go up with their stuff like I'm not there. I'm not really bold, but I always try to make eye contact.

Paul described a recent incident when he was picking up a pizza for dinner:

> We were just standing there picking it up and we were really close to the counter. The [average-height] people were pushing and pushing – violating our space. You know, I just pushed back [and said], 'Hey, you have to give us some room here.'

Such violations of personal space illustrate the ways in which everyday spaces, such as stores and restaurants, can be experienced differently by little people. Here, the amount of personal space that is acknowledged for a little person is similar to that of a child.

In addition to experiencing the spatial manifestations of dwarfism and age discourses, Paul and Mary observe ways in which gender discourses affect men and women with dwarfism differently. Mary speculated that the intersection of gender and dwarfism is related to the culturally-influenced ways in which men and women view their bodies. She noted that men may conceive of their bodies in active terms, while women may be more concerned with the attractiveness of their bodies. Hence, for men who are little people, short stature may make typically-male spaces difficult both physically and socially – especially recreational athletic facilities where physical strength and endurance are often displayed in a competitive atmosphere. As Mary explained:

> I think it is harder for a man because society thinks of the man as the taller, stronger one. And someone with dwarfism might not be able to, because of his height or physical state, do what other men can do. And a woman…it's hard for a [average-height] woman to be respected. But when you're a little person, you're really babied and looked down upon, even treated like a child.

Later in our discussion, I broached a potentially sensitive subject – the types of spaces and places that the Jamisons deliberately avoid due to fear of injury or victimization. Paul and Mary both said that they tend to shy away from large crowds. In addition to the potential awkwardness of such situations, they can even be dangerous as a little person may be shoved or knocked to the ground due their height and lower body weight. Paul noted the exception of professional baseball games, which he enjoys attending. He explained, '[When] you have your own seat…you're not in danger. I'm pretty conscious about that, about keeping out of harm's way.' In this case, segmented seating increases both the safety and enjoyment of the event. Perceptions of safety also affect the Jamison's choices of the times and places when they are in public. Although they live in a small town and feel safe in their neighbourhood, Paul and Mary feel more vulnerable to crime on the streets of larger cities, especially at night. Despite this place perception, Paul noted that they had not personally experienced any hostile incidents on the few occasions they have walked city streets at night.

Being Little in Private Space

As our lunchtime conversation ended, Paul and Mary invited me to their home where I was able to get a view of how they arrange and adapt their private space as little people. In contrast to public spaces, private spaces offer relief from inconvenience and unwanted attention. Importantly, private spaces offer the opportunity for little people to exercise more agency than they experience in public spaces. In private spaces, little people are able to find a balance between culturally-informed conceptions of how a home should be arranged and practical adaptations that make the environment more manageable. In the Jamison's home the dichotomy of public and private space was evident in terms of physical accessibility and social identity.

Both Paul and Mary were raised in households with average-height family members. As a result, their dominant discourses of home are derived from the average-size homes of their childhoods. Yet, several modifications have been made so that the couple can enjoy the ease of living that ordinary homes offer to people of average height. In the Jamison's home, spaces that are primarily functional are treated differently than spaces that contribute to the personalized milieu of their private space. Moreover, the physical arrangements of home spaces, in addition to materially facilitating everyday activities, are not separate from issues of identity (Moss and Dyck 1999).

For example, when I entered their kitchen I immediately experienced an odd sensation of spatial disorientation. The kitchen, with all the typical appliances and appointments, was scaled to the size of the Jamisons – neither Paul nor Mary appeared short in this space. Hence, in this space they appeared and seemed to feel 'normal'. The counters were approximately six inches lower than conventional counters. Consequently, I felt like a giant – completely out of proportion to the physical environment. The total *re-staturization* of their kitchen space was notable in several respects. First, it represented a considerable financial investment on their part. Paul told me that most of the little people they know are more likely to use stools for food preparation and cooking in kitchens scaled for people of average height. Also, while the modifications represent the increased agency of Paul and Mary to shape a private space, they also challenge dominant market pressures in the real estate industry. Paul mentioned that the re-staturization of the kitchen probably reduced the resale value of their home. Other more modest and less permanent adaptations were evident in their bathroom where the sink was mounted lower and clothes closets in which clothes hung at heights easily accessible to Paul and Mary.

In the living room I noticed that their family photographs and other wall decorations were hung at average height. Paul explained that he and Mary hang decorations at average height because, despite their short stature, objects look odd when hung lower. Paul's remark about the strange appearance of decorations hung lower appears to reveal that his internalized sense of 'normal' is derived from dominant discourses of what private spaces should look and feel like without regard for height. Hence, the Jamison's home helps to reinforce a 'normal' identity through its décor – even if arrangements of objects are practically inconvenient. In the less public areas of the house – the bathroom and kitchen, for example – they sacrificed 'normal' arrangements for ones of increased utility and convenience.

Starting a Family: Increased Mobility and New Identities in a Global Context

Six years after I interviewed Paul and Mary Jamison for the first time, I contacted them for a follow-up conversation. In the years between our conversations, Paul and Mary had assumed the identity of parents, and their new identity had expanded the range of spaces of their everyday lives. Paul told me that at his age (late 40s) he had not expected to be a parent. In part, it was because he had not felt a clear desire to be a father given the constraints of his everyday life related to his dwarfism. The couple, however, both enjoyed children and had worked at a summer camp for young little people for several summers. At the camp, both Paul and Mary, in addition to enjoying the kids, felt that they had something unique to offer them; they could identify with the social and physical challenges that would continue to shape the lives of the children at the camp. Although they had enjoyed sharing other people's children, Paul and Mary had never seriously entertained having biological children (Mary's condition prohibited childbirth) or adopted children (due to the complications they experience getting through everyday life). However, they began to look at their parenting prospects differently after Mary underwent surgery for the removal of cataracts that greatly improved her vision. Before the surgery, Mary had been dependent on Paul for transportation. On weekday mornings, he would drive her to work on the way to his place of employment. This arrangement had been necessary although it was time-consuming and taxing for Paul. However, one of the results of Mary's surgery was that she was able to learn to drive a car. With simple modifications to her car, including pedal extensions and a booster seat, Mary's mobility was equal to Paul's. These developments removed a barrier to the couple's mobility and caused them to revisit the idea of starting a family. Paul described the events that led up to the moment that would change their lives forever:

> We were involved with this little people camp and we'd see these kids every year and participate with them in the various activities and stuff. Then, [Mary] got her license and driving home from that camp one year I said, 'Maybe we ought to just think about adopting.' And that was like saying, 'sic 'em' to a dog! (laughs). Once I said that, as soon as she got home she was on the Internet sending out thousands of emails.

Using the Internet was not new for the Jamisons. As active members of their regional chapter of Little People of America, they used it regularly to correspond with little people across the country and to obtain information about the treatment of conditions relating to dwarfism. Just as the Internet had been instrumental in providing Paul and Mary with access to positive representations of little people and overcoming distance, it also allowed them to conduct a geographically vast search for an adoptable child with dwarfism. Paul and Mary decided to increase the probability of their finding a little person to adopt by making their search international in scope. By contacting adoption agencies with connections abroad they were able to locate several children living in orphanages in other countries. Mary explained, 'All the kids we heard about were in other countries.' I found this to be interesting and I wondered if this abundance of children with dwarfism was related to culturally-specific attitudes about the placement of little people in those societies. After months of electronic correspondence, the couple found a little girl who was living in a Russian orphanage.

Despite their eagerness to become parents, their bodies added to the bureaucratic obstacles that had to be overcome before they were allowed to assume this new identity. While home studies are part of the adoption process for every couple, the Jamisons faced additional scrutiny due to their physical conditions and short stature. For example, medical examinations and subsequent reports were deemed necessary to establish that Paul and Mary were strong enough to pick up a young child and they could physically handle the responsibilities of parenting. After they were judged as being physically fit, the Jamisons planned to travel to Russia to meet a little girl they had seen on a video sent from an orphanage in Siberia. It was clear from Paul's description that the everyday challenges of mobility are different for little people in Russia when compared to their American counterparts. He noted the inconveniences that they experienced at airports. Moreover, Paul noted a conspicuous absence of little people and other disabled people in the cities they visited.

Paul described another difference they experienced in public spaces in Russia. I asked him if Russians appeared to look at Mary and him in the same way they experience in public spaces in the United States. He explained:

> It was different than in America. They didn't stare. The Russian people were more focused on where they were going, what they were doing, and they would look and then just keep going. Whereas in America, you'll get a lot of direct stares and occasionally a rude comment. Maybe they were making rude comments in Russian, but probably not because they would just look and then continue on their way. We noticed that right away. I was okay with it. It was kind of nice. Better than being stared at and pointed at and all that.

The apparent lack of interest in the Jamisons as they walked Russian streets may be rooted in culturally-specific discourses regarding disability – instead of being viewed as a curious anomaly, the Jamisons seemed to be invisible. Another possibility is that there may be a culturally-ingrained tendency of avoiding drawing attention to one's self and others in public – a stronger sense of public privacy and of 'minding one's own business'. Whatever the cause, their seeming invisibility appeared to increase the Jamison's comfort as they met the challenges of navigating public spaces in a 'foreign' country.

The Jamison's account of the adoption is interesting in several respects. First, by choosing to adopt, they added parenting to the numerous identities, both ordinary and exceptional, that they juggle in the spaces of their everyday lives. Furthermore, by choosing to adopt a little person, Paul and Mary seemed to experience a receding of the 'disabled' identities as the mounting challenges of parenting became real to them. Hence, their experience of adopting Sarah seemed to empower them as adult little people. They had travelled distances that many average-height people never traverse, and, despite their short stature, they had successfully negotiated the challenges of international travel. If their dwarfism had increased the friction of distance in the past, Paul and Mary, as a result of their journey to adopt, seemed to reconceive physical and social distances previously defined by their stature. Also, from the account of their travels, it became clear that some of their experiences in public space that they had taken for granted throughout their lives were actually more place-contingent than they had thought – for example, their relative anonymity on streets in Russia and the accommodations required by the Americans with Disabilities Act.

Sarah, Little People and a Changing World

While I was interviewing the Jamisons at their home, Sarah told me that her favourite television programme was *Little People, Big World*. The programme is a 'reality' show that depicts the everyday life of the Roloffs, a family in which the father and mother are little people and their children are both little and average height. The family lives on a farm in Oregon and appears to be fairly well-off. Like other 'reality' shows, it purports to be a depiction of the everyday life of ordinary people – a real-time biography, rather than a scripted narrative. In actuality, Matt Roloff, the father, is a former president of Little People of America and has long been involved in advocacy and specialized products for little people. Sarah's interest in the programme offered an opportunity to discuss the impact of the show on the material and social spaces of little people.

Paul noted that there was no such programme when he was a child and that *Little People, Big World* presented positive and normal representations of little people that he thought would benefit his daughter. He thought that the show might help Sarah to see her dwarfism as more normal and less constraining than the social climate he and Mary had experienced as children. I spoke with Paul Jamison who approved of the show and posed this question: Why is *Little People, Big World* so popular *now*?

> That's an interesting question. I've actually heard some people say it just gives people permission to stare for an hour on a weekly basis. Like a freak show or voyeurism. I heard that perspective from the father of a little person. But I don't really see it that way. I see it as informative. The availability of information is so much more compared to even ten years ago. The evolution of the Internet...I guess we're more out there now. People are more aware of us, and the show is even moving that along further.

I asked Paul if the show was similar to *The Cosby Show* in the way that it challenged common discourses of minorities. He told me that *Little People, Big World*, like *The Cosby Show*, changed perceptions and was 'opening doors' for little people. He also pointed out that the show is part of a larger media phenomenon that included other television programmes, especially documentaries about dwarfism, and the increased visibility of little people on the Internet. I asked if the current presence of little people represented a new 'market' – in other words, were the changing representations of little people driven by a profit motive. Paul thought that marketing was a major factor, but did not have any objections. He explained: 'I don't have a problem with it. Some do, but as long as they're not doing something that is a poor reflection on the rest of us...Seven footers make a lot of money playing basketball. Are they exploited?' Noting that *Little People, Big World* appears to have tapped a market responsive to little people, Paul noted, with a laugh, 'Yeh, we're "in" now...it's a trend – for better or worse.' I asked if he thought that the trend of little people in the media would lead to a more permanent repositioning of little people in mainstream society. He replied: 'Nothing ever stays "in" forever, but, at least, hopefully, the awareness will be permanent. That would be good. Once the show is off the air, awareness may dwindle some, because you're not in people's living rooms.'

While the show is obviously supported by an average-size viewing public, I was curious about the effect of the portrayal of the Roloff family's everyday life on the everyday life of 'ordinary' little people. Upon hearing of plans for the show, Paul was skeptical. As he put it, 'When we first heard there was going to be a reality show based on little people, we thought, 'What kind of silliness and exploitation is this going to be?' Because most reality shows are pretty silly, we thought, 'Oh, man...'. However, Paul and Mary were pleasantly surprised with the way Matt Roloff and his family are portrayed and noted that the show was having a positive effect upon their everyday lives. They told me that the show is topical at their workplaces and often leads to conversations with average-height people that had not occurred previously. They said that some of their average height co-workers make a point of watching the show because they know little people. For some of their co-workers, the show reveals aspects of Paul's and Mary's everyday life that they were not aware of and had not thought to ask about.

They noted, with humor, the effect that *Little People, Big World* has upon Sarah, who watches the show regularly. I asked Sarah why she liked the show. 'Because they get to go places,' was her reply. Paul added, laughing, 'Yeh, she says they have a more exciting life than we do! They get to go to Disney!' Recalling an episode in which Matt Roloff constructed a glide-wire, Mary remembered seven-year old Sarah asking if they could build one in their backyard. So, it appears that, for Sarah, the depiction of active and exciting lives of little people has given her a sense of possibilities for people 'like her'. From her perspective, the show is more about having adventures than about being a little person. Clearly, the implications of such portrayals of people with dwarfism may significantly impact the socio-spatial experiences of many little people – effectively (re)placing them both physically and socially. Following up on the effect of the Roloff's adventures, I asked if there was a possibility that the show would replace one set of stereotypes with another – that it might cause viewers to over-generalize the lives of contemporary little people. Noting that the Roloffs appear to be financially well-off, Paul hoped that viewers would not think that all little people 'are rich'. I asked if average-height people might expect all men who are little people to be as outgoing, ambitious and gregarious as Matt Roloff. Paul replied:

I'd say there is some danger of that – or tendencies towards that. But people who know us know we're not like them (the Roloffs). But the Roloffs are nice folks. I mean if someone's going to think I'm like him, that's ok. He's a great example of a family man and a businessman.

Mary told me that the best thing about *Little People, Big World* is that it has 'opened up a dialogue' between little people and average-size people. She explained:

It's good for people to see who don't know a little person or who don't usually see little people. It's also good for us because I have people talking to me about it who never have spoken about the issue – because, normally, I don't discuss the issue of being little – I am just little. But now people are comfortable enough and I'm glad they are comfortable enough to talk to me about it.

Several television programmes popular in the 1960s show parallels to the current popularity of *Little People, Big World*. For example, *The Munsters* (1964-1966) and *The Addams Family* (1964-1966), like *Little People, Big World*, appeared to reflect societal anxieties regarding issues of 'difference', assimilation, and representations of the traditional family. *The Munsters* and *The Addams Family*, like *Little People, Big World*, depicted loving, energetic and enthusiastic families shown in a domestic context – one that showed them grappling with the challenges of everyday life. Much of the entertainment value of these programmes related to family members appearing 'out of place' in everyday life – especially in public space. Moreover, and somewhat disturbingly, the 'differences' exhibited in all three shows have roots in nineteenth century freak shows, when the exhibition of physical differences helped to define social space and allay the fears of the American middle class that *they* were not freakish (Thomson 1996). Could the popularity of *Little People, Big World* be intersecting with discourses that reflect cultural anxieties of displacement felt by mainstream, average-height Americans in a way similar to *The Munsters* and *The Addams Family* in the 1960s? Does the viewing of a family of little people portraying basic human emotions, hopes, frustrations help to relieve societal anxieties of feeling 'dwarfed' in an age of overwhelming information and irresolvable contradictions? Does it metaphorically express the struggle to feel at home in a world that is no longer scaled by or for 'us'? What may be a common socio-cultural experience becomes embodied and overtly expressed in the media through the physicality of the Roloff family.

Considering Geographies of Dwarfism

Considering the geographies of dwarfism offers several opportunities. First, it provides a view that broadens the literature regarding geographies of disabilities by providing personal accounts of people with dwarfism. Second, it further underscores the links between body 'difference' and socio-spatial experience noted in previous research. Similar to work on geographies of disability, gender, race, sexuality and other identities, it shows the ways in which social identities and spaces are mutually constitutive. However, beyond these similarities we can begin to understand the physical and social implications of stature in particular. While other physical characteristics are involved in the spatialization of social identities, only adults with dwarfism challenge the very scale of the built environment. While it is true that average-height children are in a similar position and, possibly, average-height adults in wheelchairs, the point here is that most public spaces are designed for adults whose body dimensions fall within a particular range deemed 'normal'. Here the relationships between space, discourse and power are explicitly manifest in the staturized regime of most public spaces. Furthermore, as the staturization of public space contributes to the potential exclusion of little people by reinforcing notions of their being 'out of place', it also involves very practical matters of safety and accessibility that shape the everyday geographies of little people.

As evident from the Jamison's home, private spaces offer the opportunity to balance practical concerns of mobility and accessibility with efforts to create a 'normal' home environment. In contrast to public spaces, little people can exercise more agency and design a hybrid home space that reflects multiple discourses of stature. This approach to domestic space is similar to the needs of other disabled groups who seek to broaden discourses of home that incorporate more ablement-sensative readings (Moss 1998). This issue is especially important for people with dwarfism because it directly impacts self-image. The fluid self-image of the Jamisons is rooted in the contrasts between public and private spaces that they experience. Hence, in the geographies of their everyday lives, the Jamisons are effectively located within what Crooks, Chouinard and Wilton (2008) refer to as 'spheres', including medical and state. Moreover, the ways in which particular places within these spheres are read and experienced influences the degree to which little people may imagine themselves as disabled or normal.

The place-contingent negotiation of their identities was complicated through the process of adopting their daughter in Russia. Through this international experience, the Jamisons learned that the spheres and places that they had experienced throughout their lives in the United States were shaped by particular discourses of disability prominent in the national culture. The Jamison's experiences in Russian cities indicate that little people's experience of public space is not ubiquitous. Rather, it is influenced by public policies that design public spaces, and particular cultural attitudes toward physical anomalies. By undergoing the rigorous examinations that qualified them as parents, the Jamisons appeared to resituate themselves in terms of their perceived 'disability'. As with other little people with whom I spoke, the Jamisons do not usually perceive themselves as especially disabled, although they encounter environments that are, in many respects, disabling. For them, disability is a fluid identity – one that is invoked, for example, when they use parking spaces specifically for disabled people, yet discarded when they interact with other parents, most of average height, in schools and other family-oriented places. Notably, recent media trends have contributed to this fluidity of the identities of little people. While there are still continuously emerging representations of little people that are degrading (such as those on Internet porn sites), the Internet has provided a format through which competing discourses of dwarfism are visible. Hence, through networking and efforts at advocacy and self-promotion, little people are able to forward normal representations of themselves in cyberspace that, they hope, will manifest in their socio-spatial experiences. Similarly, television programmes such as *Little People, Big World*, along with their related sites on the Internet, offer representations in contrast with traditions of enfreakment and offer the possibility of more inclusive and accessible conceptualizations of space.

Acknowledgements

Parts of this chapter have appeared previously in: Social Spaces of Little People: The Experiences of the Jamisons, *Social and Cultural Geography*, 2002 3(2), 175-191, reprinted by permission of Taylor and Francis Ltd, and *How Dwarfs Experience the World Around Them: The Personal Geographies of a 'Disabled' People*, Mellen Press, 2007, reprinted by permission of Edwin Mellen Press.

References

Bank, A. and Minkley, G. 1998. Genealogies of space and identity in Cape Town. *Journal of Cape History*, 25(1), 1-7.

Baxter, J. and Eyles, J. 1997. Evaluating qualitative research in social geography: establishing 'rigour' in interview analysis. *Transactions of the Institute of British Geographers*, 22, 505-525.

Bonnet, A. 2000. Whiteness in crisis. *History Today*, 50(12), 39-40.

Butler, R. 1994. Geography and vision-impaired and blind populations. *Transactions of the Institute of British Geographers*, NS 19, 366-368.

Chouinard, V. 1999. Life in the margins, in *Embodied Geogaphies*, edited by E. Teather. London: Routledge, 142-156.

Chouinard, V. 1997. Making space for disabling differences. *Environment and Planning D: Society and Space*, 15, 379-387.

Collins, D. 2005. Identity, mobility and urban place-making: exploring gay life in Manila. *Gender and Society*, 19(2), 180-198.

Crooks, V., Chouinard, V. and Wilton, R. 2008. Understanding, embracing, rejecting: women's negotiations of disability constructions and categorizations after becoming chronically ill. *Social Science and Medicine*, 67(11), 1837-1846.

England, K. 1993. Suburban pink collar ghettos: the spatial entrapment of women? *Annals of the Association of American Geographers*, 83(2), 225-242.

Gerber, D. 1996. The careers of people exhibited in freak shows, in *Freakery: Cultural Spectacles of the Extraordinary Body*, edited by R. Thomson. New York: New York University Press, 38-54.

Harvey, D. 2000. *Spaces of Hope*. Berkeley: University of California Press.

Imrie, R. 2001. Barriered and bounded places and the spatialities of disability. *Urban Studies*, 38, 231-237.

Imrie, R. 1998. Oppression, disability and access in the built environment, in *The Disability Reader*, edited by T. Shakespeare. London: Cassell, 129-146.

Kruse, R. 2007. *How Dwarfs Experience the World Around Them: The Personal Geographies of Little People*. Lewiston: Edwin Mellen Press.

Kruse, R. 2003. Narrating intersections of gender and dwarfism. *Canadian Geographer*, 47(4), 494-508.

Kruse, R. 2002. Social spaces of little people: the experiences of the Jamisons. *Social and Cultural Geography*, 3(2), 175-191.

LeCompte, M. 1987. Bias in the biography: bias and subjectivity in ethnographic research. *Anthropology and Education Quarterly*, 18(2), 43-52.

Merish, L. 1995. Cuteness and commodity aesthetics: Tom Thumb and Shirley Temple, in *Freakery: Cultural Spectacles of the Extraordinary Body*, edited by R. Thomson. New York: New York University Press, 185-206.

Moss, P. 1998. Negotiating spaces in home environments: older women living with Arthritis. *Social Science and Medicine*, 45(1), 23-33.

Moss, P. and Dyck, I. 1999. Journeying through M.E.: identity, the body and women with chronic illness, in *Embodied Geogaphies*, edited by E. Teather. London: Routledge, 157-174.

Nagar, R. 1997. Exploring methodological borderlands through oral narratives, in *Thresholds in Feminist Geography: Difference, Methodology, Representation*, edited by J. Jones, H. Nast and S. Roberts. New York: Rowman and Littlefield, 203-224.

Park, D. et al. 1998. Disability studies in human geography. *Progress in Human Geography*, 22(2), 208-233.

Pile, S. 1991. Practicing interpretative geography. *Transactions of the Institute of British Geographers*, NS 16, 458-469.

Robertson, J. 2002. Reflexivity redux: a pithy polemic on 'positionality.' *Anthropological Quarterly*, 75(4), 785-792.

Rose, G. 2001. *Visual Methodologies*. Thousand Oaks: Sage Publications.

Sibley, D. 1995. *Geographies of Exclusion*. London: Routledge.

Thomson, R. 1996. Introduction: from wonder to error – genealogy of freak discourse in modernity, in *Freakery: Cultural Spectacles of the Extraordinary Body*, edited by R. Thomson. New York: New York University Press, 1-22.

Valentine, G. 1993. (Hetero)sexing space: lesbian perceptions and experiences of everyday spaces. *Environment and Planning D: Society and Space*, 11, 295-413.

Wilton, R. and Cranford, C. 2002. Toward an understanding of the spatiality of social movements: labor organizing at a private university in Los Angeles. *Social Problems*, 49(3), 374-394.

Chapter 11

The Disabling Affects of Fat: The Emotional and Material Geographies of Some Women Who Live in Hamilton, New Zealand

Robyn Longhurst

Introduction

> By far the hardest thing for me to talk about…is that I feel my mother would love me more if I were thin…I am large and because of this I do not merit the same levels of affection and love that my thin sister does…Everything that she says, does, [and] seems to feel about me comes with a 'but' and the 'but' that she always has in her mind with regards to me is the size of my 'butt'! (Harriet).

Harriet, a participant in this research, aged in her late 20s, explains that she feels her mother does not love her wholeheartedly because she is 'large' (big, fat, out-size, plus-size, overweight, obese). Harriet had tears in her eyes as we talked. This chapter offers some thoughts on the emotional and material geographies of 12 women who identify as large or fat and who live in Hamilton, New Zealand.[1] It is argued that spaces are often disabling – emotionally and materially – for these women. A decade ago, Brendan Gleeson (1999) examined the relationship between space and disability suggesting that it is not disabilities or impairments in and of themselves that necessarily disadvantage or restrict people's lives, but rather it is social and spatial structures which take no or little account of people's embodied 'difference'. Thus, people with physical and mental disabilities are excluded from the mainstream of social activities. Others also long engaged in work on disabilities have made similar arguments. Vera Chouinard, drawing on Iris Young's (1990) research on 'cultural oppression', argues that '[f]or women with

---

1   Hamilton City, located in the central North Island in New Zealand, in 2006 had a population of 129,249 (3.2 per cent of NZ's total population). It is the fourth largest city in New Zealand and located in the Waikato region (Statistics New Zealand 2006). Hamilton has a youthful population compared to other NZ cities (median age is 31 years). In Hamilton City, 65.3 per cent of people belong to the European ethnic group, compared with 67.6 per cent for NZ as a whole. Like other Western cities, Hamilton has a thriving weight-reduction industry. The local phonebook lists 20 entries under 'Weight Reduction' and advises readers to see also 'Diet and Nutrition' and 'Dietitians'.

disabilities, negotiating spaces of everyday life, such as the home and workplace, is often a difficult, contradictory and oppressive experience' (Chouinard 1999: 142). Neither Gleeson nor Chouinard specifically mention being large or fat in their analyses but their arguments have much to offer those who want to think more about the relationship between fat bodies and spatiality.

The term 'fat' is commonly used to refer to 'corpulence, obesity, or plumpness' (Collins English Dictionary 1979: 529) but fat cannot be set apart from history, geography and culture. There are social meanings attached to being fat (see Braziel and LeBesco 2001). For example, calling someone 'fat' in contemporary Western societies is likely to cause offence because fat tends to be seen as an indicator that someone is undisciplined, unhealthy, untrustworthy and lazy. Susan Bordo (1993) traces the origins of the contemporary Western construction of the slender body as privileged over the fat body back to the Victorian era which heralded a new regime under which individuals began to deny themselves food in the pursuit of an aesthetic ideal. Fat is often thought to be unhealthy but:

> ...recent medical research shows that fat has a much more complex relationship to health than is generally acknowledged. Current investigations call into question the general threat to health posed by body fat, demonstrating that though there are some specific health risks associated with corpulence, fat is by no means the universally unhealthy condition that it is usually represented to be (Huff 2001: 47 cited in Longhurst 2005: 249).

Being fat does not simply mean being unhealthy.

Fatness is also about what Rachel Colls (2002: 219) refers to as 'emotional size', which is the way in which people perceive their bodies. Some 'average' sized or even 'thin' people (e.g., anorexics) understand themselves to be fat while some fat people understand themselves to be 'average' or perhaps even 'thin'. Body size and shape intersect with other facets of our subjectivities, such as health and gender, in complex ways. Body size is as much about how we *feel* as it is about the materiality of the physical body. How we feel might change during different life stages or even during different times of the day or night. It may depend on an array of factors such as clothing, feelings of well-being, the activity being undertaken, and interactions with people. For example, sometimes women talk about having a 'fat day' (especially when pre-menstrual). Understanding fat in this way does not mean ignoring the fleshiness of bodies but recognizing that bodies are always situated in multiple and complex material, discursive and psychological spaces.

As I have explained elsewhere,

> [o]ver the past two decades there has emerged an extensive literature by activists and academics including psychologists, anthropologists, nutritionists, sociologists and feminists on obesity, body image, diet, cultural histories of food and eating, food and sex, health, fitness, and literary and media images of fatness (Longhurst 2005: 251).

Geographers have been slower to address this topic but over the past few years some work has begun to appear. Colls (2004, 2006), for example, has reported on big women's experiences of shopping for clothes in a UK city. Colls (2007) also examined, using an 'intra-active' account of matter, the materiality of fat. Most research in geography, however, does not take this kind of socio-feminist approach (but see Bell and Valentine 1997). Rather, geographers interested in issues of population, health, mobility and public policy, not social, cultural, and feminist geographers, have been at the forefront of geographical research on fatness or 'obesity' (as it tends to be labelled in this literature).

Many have been concerned with examining 'obesogenic environments' which are all the influences (surrounds, opportunities, circumstances and conditions) that help produce fat bodies (see Swinburn and Egger 2002, Smith and Cummins 2009). Steven Cummins and Sally Macintyre (2006) provide an overview of recent findings on obesogenic environments and highlight cross-national variations in their distribution.[2] Governments and public health officials throughout the Western world have over the past few years become increasingly concerned about obesogenic environments and what is often termed an 'obesity epidemic' (Langdon 1999). Many 'experts' are now claiming that, on average, people who live in urban areas in the West are getting heavier each year. New Zealand, where I live, is no exception. The Ministry of Health in New Zealand states evidence suggests that in 2002-2003 one in three adults was overweight (and this excludes obese adults) and one in five adults was obese. These statistics are based on Body Mass Index (BMI).[3] The World Health Organization has estimated that the cost of obesity is between two and seven per cent of the annual health budget, which in New Zealand equates to approximately $303 million (see Obesity in New Zealand 2008).

By including a chapter on fatness in a book on disability I am not meaning to suggest that it is possible or useful to simply collapse fatness and disability, for example, there are interesting questions about individual agency that could be debated (see the numerous websites on this topic in relation to the law; e.g., 'Could obesity be treated as a disability?' 2008). Some people think that being fat is simply about individuals lacking the will-power to say 'no' to food and doing insufficient exercise. Others argue the relationship between body size, food intake, metabolism, genes, emotions, culture, health and place are far more

---

2   It is beyond the scope of this chapter to review this growing literature but journals such as *Health and Place*, *Social Science and Medicine*, *Environment and Planning A* and *Urban Studies* are useful places to search for work on spatial dimensions of obesity or fatness.

3   Body Mass Index (BMI) is a statistical measure of the weight of a person scaled according to height. It is defined as the individual's body weight divided by the square of their height. The formula is used widely in medicine to produce a unit of measure of kg/m^2. People are then classified as 'obese', 'overweight', 'normal weight' or 'underweight'. This approach to measuring body fat, however, is highly problematic in that it takes no account of differences in age, gender, ethnicity, fitness or muscularity.

complex (Huff 2001). Favouring the latter of these two positions, I think there is some common ground between disability and fatness in relation to spatiality that is worth thinking about (see LeBesco 2004 Chapter 6 on 'Framing fatness: popular representations of obesity as disability'). As Parr and Butler (1999: 2) suggest, there are 'many points of connection between very different people rooted in their shared experiences of biomedical inscription, pain, social isolation and political and economic marginalisation'. There are attempts to both claim and exclude fatness as a disabling condition but regardless of one's position it seems difficult to dispute that both fat people and people with disabilities are inscribed by biomedical processes, can experience social isolation, and are often constructed as 'bad' citizens who use up more than their fair share of resources (see Braziel and LeBesco 2001).

Parr and Butler (1999: 10) argue: 'The "geography of disability" can now be conceived as a diverse group of writings and materialities which encompass many aspects of embodiment and social construction'. I suggest that one such aspect of embodiment is fatness. For more than two decades geographers have been arguing that disability is an important human experience that the discipline can't ignore. So too is being fat. In this chapter I aim to build on existing work of 'geographies of fatness' by examining the experiences of 12 women who identify as large or fat and who were interviewed in-depth in 2002 in Hamilton, New Zealand. I do not address the often discussed potential health issues associated with being overweight or obese such as diabetes, heart disease, hypertension and stroke, gallstones and some cancers. The 12 women's emotional and material geographies are the focus.

Theorizing Fat Bodies

This research has been informed feminist poststructuralist geographical work on the body, disability and emotions to argue that a complex politics inhabits the bodies and spaces of fat women. Geographers studying disability have argued convincingly that people's experiences of spaces are affected by their (dis)abled corporeality and this corporeality in turn shapes the spaces that people with disabilities inhabit (Chouinard 2006, Moss and Dyck 2002 and 2003, Wilton 2003; also see Kearns and Moon 2002 on the relationship between health and place). This is also the case for people who are fat. People become inscribed as fat. This and the materiality of fatness can restrict bodily movements through space, reconstituting one's sense of self and emotional wellbeing.

I am interested in identity, discourses and emotions as well as in the materiality of bodies and the spaces in which they work, recreate, love, fight, eat, diet, exercise and so on. This theoretical framework shares much in common with that developed by Pamela Moss and Isabel Dyck (2002) to examine the lives of women with chronic illness. Moss and Dyck (2002: 9) refer to this as 'a framework for a radical body politics – one that is feminist, one that draws out discursive formation, one that has a material base, and one that is spatialized'. In keeping with this framework, I argue that certain spaces function as sites that both reinforce and contest dominant discourses about body shape and size.

The research has also been informed by work carried out over the past few years by geographers who have begun to pay more attention to 'emotional geographies' (see Bondi, Davidson and Smith 2005 and the special issue on emotional geographies in *Social and Cultural Geography* 2004). In the editorial to this special issue Joyce Davidson and Christine Milligan (2004: 523) state: 'Our emotional relations and interactions weave through and help form the fabric of our unique personal geographies. We live in worlds of pain or of pleasure; emotional environs that we sense can expand or contract in response to our experience of events.' When I collected, transcribed and read over and over again the participants' stories in 2002 I felt 'overwhelmed' by the emotions conveyed. The stories captured some of highs but especially lows that the women experienced as they attempted to negotiate a range of (dis)abling social and material spaces. The women reported often feeling excluded from the mainstream on account of their size. Examining the pleasures but mainly pains they experienced across a range of sites and scales revealed a complex geography of continual (re)negotiation with self, loved ones, family, friends and strangers that made me reflect on my own struggles with being fat. This project has undoubtedly been intimately tied to my own experiences of being 'a big-boned girl', a 'solid teenager' and a 'plus-sized woman'. This is partly why these stories have not until this point been published. The other reason is that until recently there has been little discursive space in geography to talk about emotions and affect. This is not to deny the very important work carried out by humanistic geographers in the 1970s and 1980s but there has been little work that explicitly combines emotions with critical thinking. A few years ago, however, the discipline experienced something of an 'emotional turn' (Davidson, Bondi and Smith 2005) and these kinds of 'emotional geographies' began to come to the fore as a 'legitimate' sub-disciplinary area of geographical scholarship. There is now a path to follow for those who want to write emotions geographically. Despite the delay in publishing these data, however, I think they are just as relevant today, if not even more so, given the huge media attention paid to the 'obesity epidemic' over the past few years.

Collecting and Analysing Data

I chose to focus on women for this project because numerous studies indicate that women are more likely than men to feel unhappy with their bodies and to think they are overweight (e.g., Bordo 1993, Orback 1997). The main method used to collect data was semi-structured interviews. With the help of a research assistant, Amanda, I interviewed 12 women who self-identified as fat, large, big, overweight or obese. Great care was taken when recruiting participants because we did not want to offend participants by inviting them to be part of a research project on fat bodies when they might not perceive themselves as fat. Recruitment, therefore, took place through word of mouth – snowballing – friends talking with other friends who reported they were happy to be approached. Interviews were conducted with women at a place and time of their choosing, mainly at their homes. The interviews tended to last between one and two hours, were audiotaped and transcribed in full. Participants were told that the objective of the research was to examine large or fat women's relationships with different places such as shopping malls, cafés, work places, and beaches. I explained to participants that there is already an extensive literature on body image, obesity, diet, health and fitness but human geographers have not (yet) paid much attention to this topic.

Participants were asked to start by drawing a quick sketch of themselves and to note on their sketch bits of their bodies that they like and dislike. The women were then asked approximately a dozen questions. Examples include: are there any places that you avoid because you are large and why? Are there places that you go to, but feel uncomfortable because of your size? In New Zealand there is a strong tradition of spending time at the beach. What are your experiences of being at the beach? Describe your experience of clothes shopping. Participants were not asked their weight but they were asked to fill out an information sheet that asked their age, occupation, if they have any children, living arrangements, height, and clothing size. No one objected to filling out these details and it allowed me to build a profile of the group as a whole.

The information sheets and hard copies of the transcripts were put in a folder. Detailed and repeated readings of them allowed me to identify recurring themes. For example, in nearly every interview some discussion emerged about feeling unattractive and that there were many places where the participants felt they did not 'fit' or belong. Some participants' narratives were perhaps richer than others in that these women were highly articulate. I have, therefore, in this chapter made more use of some of the narratives than others.

To aid interpretation of the interview data that follow, some information on the participants may be helpful. The women interviewed were aged between mid 20-late 60s – the average age was 44. Three identified as lesbian, the others did not explicitly identify their sexual orientation although some talked about their husbands. The women were employed in a range of occupations including lawyer, lecturer, receptionist, business systems analyst, business mentor, special education facilitator, three tertiary students, 'home executive' and two retirees. They ranged in income and social class. All were Pākehā or of European decent. This is significant because within the New Zealand context fat is a 'racialized issue'. Māori and Pacific Islanders have higher obesity rates than Pākehā and 'anti-fat' discourses tend to carry with them moral judgments about particular ethnic groups (e.g., Bell et al. 1997, Brewis et al. 1998). The smallest of the women was size 16, the largest size 30. The average size was 20-22.[4] The women ranged in height from 1.6 meters

---

4   Women's clothing sizes in NZ (and Australia) and the UK are the same. These sizes, however, tend to be one size larger than the US and Canadian sizes. For example, size 22 in NZ and the UK is equivalent to size 20 in the US and Canada.

(5 feet 2 inches) to 1.9 meters (6 feet 2 inches). They lived in a variety of different household types.

In addition to conducting interviews that generated over 200 pages of transcripts, I have reflected over a long period, on what it means to live in a body that people read as fat, flabby, large, lazy, and undisciplined. In this chapter I do not draw on the information gathered using this autobiographical method explicitly but I mention it because it has helped inform my understanding of these issues and my analysis of the interviews. In the sections that follow I focus on two main themes that emerged from the interviews: the first is 'emotional spaces'; the second is 'material spaces'.

Emotional Spaces That Matter

Liz Bondi, Joyce Davidson and Mick Smith (2005: 1 italics in original) argue 'Clearly, our emotions matter'. This statement was borne out in many of the interviews. For example, Pauline is aged in her early 20s, weighs 90 kilograms and wears a size 22. She explains that being large offers her protection:

> Because I've been attacked and raped, and I was constantly under attack when I was like that [slim] and I don't get that anymore which is comforting in its own way, but I do understand this [fat] as a protection. I put a wall around me…At the moment I'm all fuddy-duddy, someone who waddles around and giggles and smiles and I'm not taken on that level of being a hot chick who is to be approached.

There is a complex gendered politics that surrounds body size, image, perception, abuse, violence and rape. It is not uncommon for sexual abuse survivors to develop a hatred of their bodies. Pauline describes herself in unbecoming terms – as someone who is 'all fuddy-duddy' and who 'waddles around'. In her sketch Pauline notes that she likes her face and her breasts have potential 'as long as they have much support' but she says she is 'ashamed of her love handles and cellulite', has 'legs like a bloody Clydesdale [horse]', and 'can't stand' and 'can't show anybody' her stomach (see Figure 11.1).

Pauline is not alone in loathing her body and expecting that others will also loathe it. This was evident in all but two of the 12 participants' accounts of their corporeality (although none of the other participants discussed having been abused or violated). Sandra, aged in her mid 50s, describes herself in the interview as 'a big thing'. She explains that she enjoys dancing but questions 'who wants to dance with a big thing?' Sandra's description of herself resonates with Julia Kristeva's concept of the abject as described in *Powers of Horror*. Kristeva (1982) offers an account of the psychic origins and mechanisms of disgust that help explain a subject's loss of distinction between themselves and objects/others. In referring to herself as a 'big thing' Sandra unsettles the self and the conscious by disrupting the border between being and non-being, between the human and the 'thing' (see Wilton 2003).

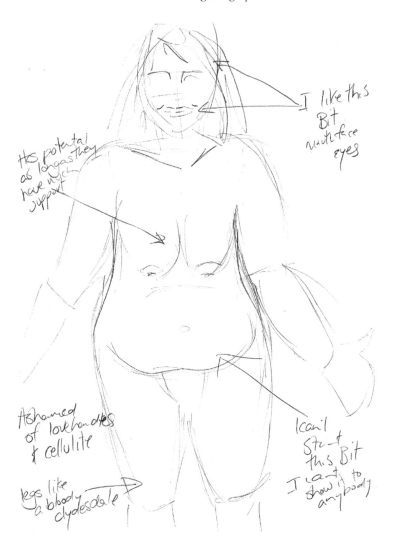

Its potential
as long as they
have neck
support

I like this
Bit
mouth face
eyes

Ashamed
of love handles
& cellulite

legs like
a bloody
clydesdale

I can't
Stand
this Bit
I can't to
show anybody

Figure 11.1   Pauline's self-portrait

Jan, aged 30, cares for her three-year-old son. She wears a size 16 and was one of the smallest participants in the study. Despite not being very large she explains that she wouldn't go to the beach because she thinks she might 'gross people out' by 'bulging out the sides' of her swimsuit. Jan says:

> it's not really a nice look…it's a little bit like looking at a person who's got a bit of food hanging in between their teeth…you can't say anything because you don't want to embarrass them and so you end up looking away and that's sort

of how I feel about being fat, you know, everyone actually wants to say 'do you know that your ass is actually hanging out of your bikini?' but everyone is too embarrassed to say.

Jan expresses concern over how others might judge her body. She says: 'I often get the feeling that because of my weight people make a judgment call about the fact that I may be lazy, maybe have no self-discipline…even though people say you can't judge a book by its cover, we all do it.'

Fat people are stereotypically constructed as lazy, inadequate, lacking in self-discipline, not to be trusted and unhealthy (see Bell and Valentine 1997).

Denise is not overly concerned with how others might judge her but this doesn't mean she does not face challenges in her daily life. She is 1.9 metres (6 feet 2 inches) tall. It is difficult for Denise to find shoes that fit her large feet. She also finds shopping for clothes problematic. Denise says: 'I just feel stupid. They are all size 8 and I am size 20. Then if you go to the big women's clothes they cater for women that are my height but really really big so you get the tent dresses'. Denise avoids shopping for clothes as much as possible. She explains 'I can't even remember how much I weigh because I don't weigh myself much'. Interestingly, Denise then explains:

I don't exist as a woman a lot of the time. It's very weird, very very odd. I used to walk around at night when I was younger and not worry about it because I thought people would think I was a man at night. But it was dangerous. I've had dangerous things happen…so it's not safe but you delude yourself when people think that you are big but it's dumb. I shouldn't do it. Anyway, I don't do it now.

Denise says that she also doesn't go to pubs very often because she gets 'hassled'. She explains: 'I just don't like being treated like crap so I don't go…I don't want people harassing me or treating me like a piece of crap because I am female and I'm tall'. Denise also avoids beaches explaining: 'I don't wear togs [swimsuit]. I don't wear them anymore. I used to but I don't now because I feel too big…it's the thin thing, the thin thing is everywhere, isn't it? It's not just at the beach'. Denise's interview offers some interesting insights into the disabling affects of being a 'big' woman – tall and large. On the one hand, there are many spaces in which, for many years, she has felt uncomfortable and unwelcome. These include women's clothes shops, pubs and beaches. On the other hand, Denise has resisted being spatially confined. When younger she used to 'walk around at night and not worry' because people would think she was a man. Spatial confinement is not the whole story of Denise's life. Denise resists the 'cult of thinness' (Hesse-Biber 1996), even if today that resistance is somewhat more covert than it used to be. Denise says she no longer 'battles' with people who give her a 'hard time' about her size and that it is easier to walk away.

I'd rather hang out with people who just don't worry about things like that...it's
not important. I've worked with a lot of disabled people and a lot of elderly
people who can't move so why am I gonna freak out because I'm big. It's not
my policy – that perfect size 10 and petite.

Approximately half of the participants also resisted feeling 'weighed down' by
expectations to conform to the 'cult of thinness'. Miranda, who is 50 and works
as a receptionist, says that probably one of her biggest regrets is that for years
she did not go to the beach. Last year, however, she thought 'to hell with it'
and brought herself a swimsuit. Miranda says: 'I didn't care. I thought people
can look at me and think "she's gross", and she is [I am], but Allen [husband]
thought I was alright and that's all that I worried about'. For Miranda, the beach
became her 'geography of resistance'. Steve Pile and Michael Keith (1997)
argue that resistance is not simply the binary opposite of domination, nor is it
something that can be plucked from spatiality. It is complex, contradictory and
often paradoxical. There are 'geographies of resistance'.

Given the contradictory nature of resistance it is perhaps not surprising that
approximately half of the participants seemed to recognize and articulate both
the dominant discourses and reverse discourses that surround body size. At times
they reiterated the dominant discourse making disparaging comments about their
bodies, at other times they claimed to not care about being fat. Pauline, a university
student, reflects on this:

All these competing discourses...things going on in my head 'I'm a strong
woman I don't need to get thin' I have just as much right as you. You're a fat
ugly bitch, no one is gonna love you like this. Things would be better if you lose
weight. Everybody else manages to do this. Just all of these different thoughts
going on and they're going ping ping ping, all over the place all of the time.

There is much debate on what are the most effective ways to resist and transgress
societal relations, processes, and/or institutions that foster exploitation and
domination. Fat people have long devised strategies for resisting the 'tyranny of
slenderness' (Chernin 1983; also see LeBesco 2004 for accounts of the size acceptance
movement in the US and Bovey 2002 in the UK). There have emerged fat acceptance
groups and magazines that celebrate largeness. A growing number of Internet sites
(including personal advertisements, fashion and sex sites) are catering for fat people
and their admirers. Designers and retailers have begun to discover the 'plus-size'
fashion market. Stores that promote and sell glamorous clothing for fat women, such
as Precious Vessels owned by Cindy Gibbons (see Gibbons 2002) in New Zealand
and Anna Scholz's design stores in London, can be read as sites of resistance (but
also domination in that they promote hegemonic constructions of femininity).

One participant in this study actively contested both slimness and hegemonic
constructions of heterosexual femininity. Tam, aged in her mid 40s, buys men's

clothes with a number of Xs before the L[arge]. When I asked her if there are any places that she avoids on account of her size, Tam replies:

> Not really. I don't give a fuck most of the time if people think I'm fat but I have felt uncomfortable in changing rooms 'cause it always feels weird trying on clothes in such a small confined place. I feel more uncomfortable in certain clothing than in places, like straight women's clothing isn't made for fat chicks. I prefer men's gear 'cause I'm butch and 'cause it's more comfortable. I think when I get loud I notice that I am bigger than other people 'cause it's when I feel like I take up too much space but on the whole I like taking up space. Hey, it's there, [I'll] fill it.

Tam does not feel that she needs to shrink from spaces but that she can fill them and be present in the world as someone who is proud (see Probyn 2005 on the politics of pride) of being fat and butch. She actively resists constructions of the 'slim', heterosexual woman who men supposedly find sexually desirable. Slimness, as a cultural fixation, especially for women who want to appear attractive to men, has been critiqued in recent years (Bordo 1993). Research has been carried out on sex, excess and pleasure (Kent, 2001 on FaT GiRL), fat pornography (Braziel 2001 on 'Plumpers and big women'), the carnivalesque (see Stukator 2001), and fat beauty (Klein 2001). Tam's narrative illustrates some of the possibilities for thinking about queer geographies of fatness. Some argue that in the first instance claiming, accepting and using the term 'fat' can be seen as a resistance of sorts and can help individuals to get on and lead positive lives regardless of body size, instead of forever wanting and waiting to be slim. It is a way of coping with and resisting the current stigmatization of fatness.

Emotional geographies are lived through bodies. Like all people, the women who took part in this research experience a range of complex emotions every day (joy, pleasure, anger and disappointment to name just a few), however, when thinking about their body size and shape they were far more likely to express negative rather than positive feelings. When I invited the participants to sketch themselves I asked that they note things that they both like and dislike about their bodies, but few wrote positive things. Instead phrases such as 'hate tummy', 'tummy flap', 'sweaty under boobs' and 'chubby hands' were typical. At least half of the participants expressed some disgust towards their own bodies. Elizabeth Grosz (1994: 195) suggests that fat, fleshy, soft bodies disrupt the solidity of things and that in this way they can be seen to occupy a borderline state that disturbs identity, system and order. Colls (2007: 358) argues that because fat bodies disturb the 'norm' they are discursively positioned as 'out of place' and/or 'in need of reduction'.

The interviews revealed that the 12 women faced numerous emotional challenges living in a fat-phobic society. However, they face not just emotional challenges but also material challenges trying to 'fit' into physical spaces that have not been designed for their corporeality. As is the case with many people with disabilities, there is little attempt to accommodate their bodily 'difference'. As will be seen in the section that follows many of the participants commonly adopted strategies of avoiding visiting certain places, sitting in particular seats, wearing certain items of clothing, or partaking in activities that meant exposing or using their bodies in particular ways. Avoiding places and situations was a way of dealing with the fear of being constructed as Other. This lack of accommodation of large bodies can be read as an attempt to confine them because of the threat they pose to a public world intent on whittling down the space they are permitted to occupy.

Material Spaces That Don't Fit

People often experience confinement, that is, a restriction and shrinking of their (emotional, discursive and physical) horizons as they grow fatter. Confinement is a recurring theme in fat people's lives. They are constantly confined, defined and delimited by others. Perhaps it is not surprising, then, that Denise who is both tall and large desires 'big spaces'. She explains:

> In direct relation to my big size, I want big spaces. I want a big house. I would like to design a house with a tall toilet I would like a tall toilet and my friends can have a step and I am quite happy with that...I need lots of space. It's not my idea of fun being in small spaces.

When I asked Sandra if there were places where she felt that she did not fit she replied: 'I don't fit well into airplane seats anymore, or hairdressers. I used to go to a salon but I couldn't fit into their waiting seats. Coffee lounges, some of their seats are too small'. Miranda, aged in her mid 40s, feels self-conscious at the rugby (watching rugby is an important recreation for many New Zealanders). She says: 'you are sitting in allocated seats. They are all joined together and they are really little, they are actually tiny'. Miranda says another place she doesn't like going is the local theatre: 'I hate going there. The seats are really tiny...I feel like I'm really exploding on to someone else's seat...I am really conscious of that'. Linda McDowell (1999) makes the point that women are often more conscious than men of encroaching on the space of others. She comments: 'As most women know, especially if they have recently made a plane journey in economy class, seated next to almost any man, men seem to feel that they are entitled to occupy more of the available space than women' (McDowell 1999: 41).

Harriet was aware of encroaching on what is considered to be other people's space, even to the point of feeling that she needs to lie when checking in for a flight at the airport in an attempt to secure more space not just for herself but also for those around her. Harriet says:

> I feel uncomfortable and have a heightened sense of my size in places where the space is small...like on planes...I sort of think anyone I have to sit next to will see me coming and think 'oh god, I don't want to be squashed by that'. I always

sit in an aisle seat…if I'm too late checking in to get an aisle seat automatically I will lie and say that I get claustrophobia and so need to be on an aisle.

For Pauline, aged in her early 20s, getting out of lecture theatre seats is difficult. She says: 'I don't know if it's just me or badly designed places…If I'm feeling good I'll curse the seat, if I'm feeling bad I'll curse myself'. At least five of the participants mentioned avoiding some seating arrangements (such as fixed seating in restaurants) or particular chairs. Heather, aged in her early 40s, says she feels confined in restaurants that have built-in seating:

> I felt hesitant when we went out to [name of restaurant] the other night for dinner and I got there and thought 'God, I hope they don't put us in one of those booths' because I know I can fit in the booth but I never feel comfortable…[it's the] same with the fixed seating at places like McDonalds.

It is not only seating in public spaces that some large people think carefully about but also seating in their own homes and workplaces. Heather continues:

> If we have got other people around then I wouldn't sit on the two-seater [couch] with them. I would prefer to sit where we have got just one chair…a separate chair so I don't have to sit next to them. Our outdoor furniture is quite big, so I fit in that [but] I know that I have in the past felt uncomfortable in plastic [outdoor] chairs.

At least three participants were anxious about whether particular chairs or stools were capable of supporting their body weight. Falling through a chair, or having a chair collapse under you, was seen by these participants as being the ultimate in humiliation.

Sandra likes to engage in outdoor adventure activities including white-water rafting. Sometimes, however, this proves to be humiliating when the clothing that tour operators supply is too small. Sandra says she asked the white-water tour operator if their wet suits were large big enough for her. Despite their assurances, she finds that 'Often they are not big enough and it's really embarrassing. It is exhausting. I actually use all my energy trying to get into the suits'. Sandra was heartened though when she went jet-boating and they had an overall and coat that were large enough. She was able to put these on without embarrassment.

Unlike Sandra, Jan who is aged in her late 20s, avoids going to places or partaking in activities that might result in embarrassing situations, even in her own home. Jan says:

> Beaches, changing rooms, you know, open plan changing rooms…walking around the house naked. Yeah, I wouldn't do that now. Ronnie's [her husband's] mum has got a massive mirror…I got such a shock the first night. I thought who is that woman because I just had my underwear on, like shit, yeah, that was a real shock!

Jan is 1.6 metres (5 foot, 4 inches) and size 16. She is not very large but still feels beaches, swimming pools, changing rooms, and places with mirrors are all to be avoided.

Height was a significant factor for several of the participants (those who felt they were outside 'the norm'). It intersected with body mass in different and sometimes contradictory ways. For example, Jan felt large partly because she perceived herself to be short. Denise, at over six feet tall, felt large partly because she perceived herself to be tall. Denise describes walking into the changing room in a women's clothing store as being like walking into a dolls' house.

> They're [changing rooms] never really tall enough. Last time I went to [name of shop] I had to crouch down. It's ridiculous. People could see my boobs...I think it's just crazy. I mean I'm not the only person who is tall....Why is everything so small? Let's walk in to a doll's house and try and cram these clothes on and everyone can see us – in fact why don't we just do it in the street?

Denise's anger, expressed in the sarcastic remark 'why don't we just do it [get changed] in the street?' is born out of years of frustration of feeling that she doesn't fit comfortably in her surroundings.

Tam, like Denise, rallies against the many spaces that feel too small for her. I asked Tam if she ever felt that spaces are too small and she replied with an emphatic 'Shit yes. Women's toilets are ridiculously small, movie theatre seats are just an insult and Japanese cars just don't let big butch dykes feel comfortable. It's like these places are only for twigs.' Most of the participants feel guilty about taking up what they perceive to be too much space but Tam resists feeling marginalized in this way representing her own body as the norm and smaller bodies that fit into these spaces as 'twigs'. She contests processes of 'cultural oppression' (Young 1990) and proudly takes up space instead of trying to limit herself within particular discursive and material confines.

Conclusion

Policy makers, practitioners, activists and academics in a range of disciplines are now addressing the issue of fat. Social, cultural, health and disabilities geographers are well positioned to make contributions in this area by considering the relationship between fat subjectivity, disability and spatiality. I am not arguing that fatness is necessarily a disability nor that these two axes of subjectivity are mutually exclusive. What is evident, however, is that environments, both emotional and material, can be disabling for fat people, especially for fat women who face societal pressure to conform to slimness as a gendered norm. Having a disability is often considered to be outside of an individual's control – the result of an accident or birth 'defect' perhaps – but fatness tends to be seen as the product of an individual's appetite, greed and indolence. It is commonly thought that people ought to be able to control their weight. To not control one's weight is thought to be irresponsible and a drain on the public purse. In this way, there are important differences between people who are fat and people who have a disability, illness or impairment but nevertheless there are some similarities.

Both fat people and people with disabilities are oppressed by dominant power relations that revere slimness and able-bodiedness (whether that is physical and/or mental). Both are forced to negotiate disabling power structures and spaces in their everyday lives. Like 'disabled bodies', fat bodies are seen to be 'dissident bodies', meaning they are 'particularly resistant to articulated norms' (Dorn 1994: 154). Chouinard (1997, 1999, 2006) for many years has argued that people with disabilities are constructed as 'out of place' and forced to inhabit 'ableist spaces'. She says: 'Coming to terms with being constructed as negatively different as a result of disability often involves periods of personal crisis, transition and growth that are both painful and liberating' (Chouinard 1999: 143). Not everyone who is fat considers themselves to be disabled but many people who are fat do experience having to come to terms with 'being constructed as negatively different' in a society that venerates slimness.

In this chapter I have been able to do little more than make a cursory start on exploring some of the complex intersections between fatness and disability. Both are discursively constructed as bodily 'deficiencies' – 'conditions' that are undesirable and best rectified in pursuit of a 'better' body, or at least one that conforms to normative ideals. Both have a fleshy materiality that cannot be ignored. The participants in this study were not asked directly whether they felt or experienced their size as a disability, illness or impairment instead this emerged as an important theme in the research. Future work is required to explore in more depth the complex intersections between the materiality and emotionality of being fat and/or disabled.

Acknowledgements

I would like to thank the women who took part in this research. I feel humble and enormously appreciative that you trusted me with your stories. Amanda Banks provided useful research assistance for which I am grateful. Finally, thanks must go to the editors, Vera Chouinard, Robert Wilton and Edward Hall for initiating and seeing through to completion this book project.

References

Aaronovitch, D. 2000. Fat is not only a feminist issue to us larger men. *The Independent*, London, 9 June.
Bell, A., Amosa, H. and Swinburn, B. 1997. Nutrition knowledge and practices of Samoans in Auckland. *Pacific Health Dialog*, 4(2), 26-32.

Bell, D. and Valentine, G. 1997. *Consuming Geographies*. London: Routledge.

Bondi, L., Davidson, J. and Smith, S. 2005. Introduction: geography's 'emotional turn', in *Emotional Geographies*, edited by J. Davidson, L. Bondi and S. Smith. Aldershot: Ashgate,1-16.

Bordo, S. 1993. *Unbearable Weight: Feminism, Western Culture, and the Body*. Berkeley: University of California Press.

Bovey, S. 2002. *What Have You Got to Lose? The Great Weight Debate and How to Diet Successfully*. The Women's Press, London.

Braziel, J.E. 2001. Sex and fat chics: deterritorializing the fat female body, in *Bodies Out of Bounds: Fatness and Transgression*, edited by J.E. Braziel and K. LeBesco. Berkeley: University of California Press, 231-254.

Braziel, J.E. and LeBesco, K. (eds) 2001. *Bodies Out of Bounds: Fatness and Transgression*. Berkeley: University of California Press.

Brewis, A., McGarvey, S., Jones, J. and Swinburn, B. 1998. Perceptions of body size in Pacific Islanders, *International Journal of Obesity*, 22, 185-189.

Chernin, K. 1983. *Womansize: the Tyranny of Slenderness*. London: The Women's Press.

Chouinard, V. 1997. Making space for disabling differences: challenging ableist geographies (guest editorial essay). *Environment and Planning D: Society and Space*, 15(4), 379-386.

Chouinard, V. 1999. Being out of place: disabled women's explorations of ableist spaces in *Embodied Geographies: Spaces, Bodies and Rites of Passage* edited by E. Teather. London and New York: Routledge, 142-156.

Chouinard, V. 2006. On the dialectics of differencing: disabled women, the state and housing issues. *Gender, Place and Culture*, 13(4), 401-417.

Collins English Dictionary 1979. London, Williams Sons Collins and Co., p. 316.

Colls, R. 2002. Review of 'Bodies out of bounds: fatness and transgression'. *Gender, Place and Culture: A Journal of Feminist Geography*, 9, 218-220.

Colls, R. 2004. 'Looking alright, feeling alright': emotions, sizing and the geographies of women's experiences of clothing consumption. *Social and Cultural Geography*, 5, 583-596.

Colls, R. 2006. Outsize/Outside: bodily bignesses and the emotional experiences of British women shopping for clothes. *Gender, Place and Culture*, 13(5), 529-545.

Colls, R. 2007. Materialising bodily matter: intra-action and the embodiment of 'fat', *Geoforum*, 38, 353-365.

Could obesity be treated as a disability? 2008. Available at: http://blogs.nzherald.co.nz/blog/all-days-work/2008/6/25/could-obesity-be-treated-disability/ [accessed: 9 September 2008].

Craven, E. 1996. Readings of the obese body: embodied geographies, in *All Over the Place. Proceedings of a Postgraduate Conference in Social and Cultural Geography.* Compiled by P. Shurmer-Smith. University of Portsmouth. Working Paper in Geography No. 26, 29-36 [available from the Department of Geography, University of Portsmouth].

Cummins, S. and Macintyre, S. 2006. Food environments and obesity – neighbourhood or nation? *International Journal of Epidemiology*, 3(1), 100-104.

Davidson, J., Bondi, L. and Smith, S. (eds) 2005. *Emotional Geographies*. Aldershot: Ashgate.

Davidson, J. and Milligan, C. 2004. Editorial: embodying emotion sensing space: introducing emotional geographies, *Social and Cultural Geography*, 5(4), 523-532.

Dorn, M. 1994. Disability as spatial dissidence: a cultural geography of the stigmatized body, unpublished MSc thesis, The Pennsylvania State University.

Gibbons, C. 2002. *Breaking the Cycle*. Auckland: Random House.

Gleeson, B. 1999. *Geographies of Disability*. London: Routledge.

Grosz, E. 1994. *Volatile Bodies: Toward a Corporeal Feminism*. St Leonards: Allen and Unwin.

Hesse-Biber, S. 1996. *Am I Thin Enough Yet?* New York: Oxford University Press.

Hopwood, C. 1995. My discourse/my-self: therapy as possibility (for women who eat compulsively). *Feminist Review*, Spring, 49, 66-82.

Huff, J.L. 2001. A horror of corpulence: interrogating Bantingism and mid-nineteenth-century fat-phobia, in *Bodies out of Bounds: Fatness and Transgression*, edited by J.E. Braziel and K. LeBesco. Berkeley: University of California Press, 39-59.

Kearns, R. and Moon, G. 2002. From medical to health geography: novelty, place and theory after a decade of change'. *Progress in Human Geography*, 26(5), 605-625.

Kent, Le'a. 2001. Fighting abjection: representing fat women, in *Bodies out of Bounds: Fatness and Transgression*, edited by J.E. Braziel and K. LeBesco. Berkeley: University of California Press, 130-151.

Klein, R. 2001: Fat beauty, in *Bodies out of Bounds: Fatness and Transgression*, edited by J.E. Braziel and K. LeBesco. Berkeley: University of California Press, 19-38.

Kristeva, J. 1982. *The Powers of Horror: An Essay on Abjection*. Trans. Leon. S. Roudiez. New York: Columbia University Press.

Langdon, C. 1999. We're all getting too fat, survey shows. *The Dominion*, 26 August, Edition 2, 3.

LeBesco, K. 2004. *Revolting Bodies? The Struggle to Redefine Fat Identity*. Massachusetts: University of Massachusetts Press.

Longhurst, R. 2005. Fat bodies: developing geographical research agendas. *Progress in Human Geography*, 29(3), 247-259.

McDowell, L. 1999. *Gender, Identity and Place: Understanding Feminist Geographies*. Cambridge: Polity with Blackwell.

Moss, P. and Dyck, I. 2002. *Women, Body, Illness: Space and Identity in the Everyday Lives of Women with Chronic Illness*. Rowman and Littlefield, Lanham.

Moss, P. and Dyck, I. 2003. Embodying social geography, in *Handbook of Cultural Geography*, edited by K. Anderson, M. Domosh, S. Pile and N. Thrift. London: Sage, 58-73.

Obesity in New Zealand 2008. Available at: http://www.moh.govt.nz/obesity [accessed 30 July 2008].

Orbach, S. 1997. *Fat is a Feminist Issue: the Anti-Diet Guide for Women*. New York: Galahad Books.

Parr, H. and Butler, R. 1999. New geographies of illness, impairment and disability, in *Mind and Body Spaces: Geographies of Illness, Impairment and Disability*, edited by R. Butler and H. Parr. London: Routledge, 1-24.

Pile, S. and Keith, M. (eds) 1997. *Geographies of Resistance*. London, Routledge.

Probyn, E. 2005. *Blush: Faces of Shame*, Minneapolis: University of Minnesota Press.

Procter, K.L., Clarke, G.P., Ransley, J.K. and Cade, J. 2008. Micro-level analysis of childhood obesity, diet, physical activity, residential socioeconomic and social capital variables: where are the obesogenic environments in Leeds? *Area*, 40(3), 323-340.

Smith, D.M. and Cummins, S. 2009. Obese cities: how our environment shapes overweight. *Geography Compass*, 31(1), 518-535.

Statistics New Zealand 2006. Available at http://www.stats.govt.nz/NR/rdonlyres/5E6FB3C6-7F3D-416E-88A9-9C4993FD3BFA/0/HamiltonCity.pdf [accessed 11 September 2008].

Stukator, A. 2001. 'It's not over until the fat lady sings': comedy, the carnivalesque, and body politics, in *Bodies Out of Bounds: Fatness and Transgression,* edited by J.E. Braziel and K. LeBesco. Berkeley: University of California Press, 197-213.

Swinburn, B. and Egger, G. 2002. Preventive strategies against weight gain and obesity. *Obesity Reviews*, 3(4), 289-301.

Wilton, R.D. 2003. Locating physical disability in Freudian and Lacanian psychoanalysis: problems and prospects. *Social and Cultural Geography*, 4(3), 369-389.

Young, I.M. 1990. *Justice and the Politics of Difference*. Princeton: New Jersey: Princeton University Press.

Chapter 12

Embodied Ageing in Place:
What Does it Mean to Grow Old?

Janine L. Wiles and Ruth E.S. Allen

Introduction

Ageing is a form of social and physical difference. In this chapter, we connect thinking on ageing with thinking on disabilities to argue that theoretical developments in geographical disability studies have already and can further inform our understanding of the spaces in and through which the 'difference' of ageing is produced, embodied, and experienced, and vice versa.

Though a significant proportion of 'people with disabilities' are older people, we are not defining 'old age' in and of itself as a disability. We do argue however, that pervasive negative social discourses about old age are a form of disabling geography, in that they devalue old age and normalise negative associations with old age. This in turn allows for the failure to account for the physical and social environmental needs and interactions of older people. Physically disabling environments (such as homes poorly designed, uneven footpaths, inadequate public transport) and socially hostile attitudes towards old age (devaluing old age, or defining successful ageing on narrow criteria) are all examples of disabling geographies.

Ageing is a fundamentally human aspect of being. Each of us is growing older, yet very few people will admit to being 'old'. Simone de Beauvoir (1970) observed ambivalence towards older adults verging on 'biological repugnance'. She argued that intergenerational exchanges are marked by duplicity; formal deference underlain by resentment, depreciation and manipulation of the older person. This cynical view of old age, and the oppressive power relations that surround it, permeates widely held views of ageing. Yet these are counterbalanced by strong social repertoires of reverence and respect for older people, and growing discourses on positive or successful ageing.

In this chapter, we highlight recent theoretical debates around ageing, and suggest some important points of similarity between this work and disability scholarship. We suggest that there are at least four ways in which disability scholarship can help gerontological researchers understand ambivalence around ageing, and the potential impact of different ways of understanding the ageing process. First, like categories of 'disability', there is reluctance to embrace the identity of 'old age' and that of 'older person', whether because people do not see this identity as part of who they are, or because they recognise it as a devalued and stigmatised category.

Second, the trajectory of the debate on ageing echoes that of disability, from clinically defined 'problems', to disability as a social construct, to disability as embodied difference, as mapped out by disability theorists. While dominant medical approaches still seek to 'treat' both old age and disability, critical social theorists have sought ways to challenge, reconstruct, or provide alternatives to such ways of understanding both disability and old age.

A third area of similarity is the need to address the difficult tension between the urgency of addressing the inequities and injustices associated with the way that either old age or disability is defined, versus the importance of recognising, supporting, and celebrating difference and diversity. For both gerontological and disability theorists, there is much to be learned from connections to debates around sexism and racism and the tension between fighting for equity as opposed to struggling for equality.

A fourth point of contact is around the issue of language. For those interested in either disability or gerontological issues, language is power. The difficulty of naming, of creating an inclusive vocabulary that is meaningful rather than cosmetic or euphemistic, and that recognises difference without relegating it to a somehow lower category, is central to both. Language has the ability to reproduce the status quo as well as to challenge it, to shape as well as to reflect attitudes towards difference, to focus our thinking towards one explanation without considering wider possibilities.

In addition to outlining some of the theoretical debates of interest in the literature, we will explore briefly some thoughts of older people themselves who have participated in some of our research. These 'experts' are often the least consulted on matters concerning these issues, a situation unfortunately also evident in at least some disability scholarship.

Theories on Ageing

There are a wide variety of ways that researchers have understood the phenomenon of ageing, models that at times contradict each other or ignore areas that other models or theories prioritise. Some key ideas are summarised in Table 12.1, in order to give an overview; key theories will be explored in relation to the four points of overlap with disability scholarship.

Ambivalence about 'Older' Identity

The reluctance to embrace the identity of 'older person' may be similar to the ambivalence around being categorised as 'being disabled' (Crooks, Chouinard and Wilton 2008, Wendell 1996). Ambivalence operates at a personal level, where the 'older' identity is not seen as part of who one is, as research entitled 'We're not old!' (Hurd 1999) highlights. At a broader social level, 'oldness' is an identity that is socially stigmatised and devalued.

Many researchers have been struck by the strong theme of ambivalence when people talk about age and getting old. For example, in biographical interviews with people aged 90 years and older, many said they were not old (Jolanki, Jylhä and Hervonen 2000). Similarly, Rebecca Jones (2006) found older people sometimes talk as if they are old, and at other times as if they are not, often choosing to reject the category of age because of the negative values associated with older age. Most of what people define as happening in old age are negative changes and deterioration, with very few positive ideas (see Table 12.2).

What Do Older People Say?

In a study which involved extended conversations about sense of place with 83 community-dwelling older people living in Auckland, New Zealand (Wiles et al. 2009), we specifically asked participants to tell us, 'What is good about growing older?' They were all aged over 75 years, average age 80 years, with 14 participants over 85. The question was deliberately framed to include the positive emphasis, we hoped to elicit views other than the dominant notion of ageing as a negative experience. Many were surprised by the question and about a quarter said that 'Nothing' was good about it. Of these, five mentioned physical deterioration (aches and pains, leg trouble, reduced hearing and vision, less energy, 'body breaks down'), and associated loss of independence or confidence. In the 'nothing good' group there was also a sense of powerlessness, as in the comment 'I didn't know what to expect' or the rather fruitless exhortations, 'Don't grow old' and 'I wouldn't advise it.' The impact of the dominant biomedical views on ageing, emphasising physical deterioration and a concomitant desire to 'treat' or do away with old age, were evident in their talk. Comments from the majority of participants at odds with these negative views will be explored further below.

The inherent ageism in even discussing a rejection of 'oldness' is also interesting. Initial debate as to whether ageism exists (Kite et al. 2005) has now been replaced by calls to examine the negative consequences of ageism in everyday life, from transport and toilet facilities to health and social service provision. There is an expectation that the 'baby boomer' generation may be less accommodating of age prejudice than current cohorts (Minichiello, Browne and Kendig 2000) and may be able to draw lessons from the disability movement in acting against 'disabling' social conditions that turn impairments or loss of function into 'disability' (Oldman 2002). But the pervasiveness of negative stereotypes and ambivalence around being 'old' mitigate against such activism; in order to fight against ageism, there needs to be some ownership of the identity of 'old'.

Table 12.1   Theorising old age

	Domain of interest	Some key ideas	Some key models/theories	Critiques
Primarily individual focus	*Ageing identity is related to biological, physiological decline*			
	The body	Biomedical view. Cellular and organism decline; disease prevalence increases with age. Seeks ways to more effectively prevent, diagnose and treat diseases of old age	Dominant model in medical and health science endeavour; technology and pharmaceutical development; influences health and social policy, re 'burden' of sick old people	Focus on bodies/body parts as medical 'problems' with medical solutions; though impact of social/economic determinants on health/ageing increasingly being acknowledged
	Ageing identity is related to activities and roles			
↕	The appropriate roles, activities and developmental tasks that older people can engage in to maximise well-being	Seeks to find 'successful' or 'appropriate' ways of ageing, in terms of roles and activities, levels of engagement or withdrawal from society; achieving certain individual/family developmental tasks in a timely way; achieving key health or social behaviours needed to maintain well-being	Disengagement theory Activity theory Successful ageing Selective Optimisation with Compensation (SOC) Developmental stages/tasks of ageing	Who defines 'success' or 'failure'? Often normative ideas reflecting social mores of the time; emphasis on individual responsibility regardless of context may 'blame the victim' if health behaviours weren't maintained or developmental tasks achieved
Primarily socio-cultural focus	*Ageing identity is related to broader social structures and cultural norms*			
	How personal experiences of ageing are the result of wider social, cultural and political practices	Seeks to understand how wider socio-cultural influences can cause the 'problems' of old age, with structures like retirement excluding older people and impoverishing them; power issues of social difference (gender, class etc.) and ageism; how negative discourses constitute ageing	Political economist Feminist gerontology Foucauldian gerontology Discursive theorists	Leaves out the material body – physical pain or disease is not just socially constructed. Can homogenise older people as hapless victims of dark social powers

Table 12.1 continued   Theorising old age

	Domain of interest	Some key ideas	Some key models/theories	Critiques
Reclaiming the body	*Ageing identity is embodied*			
	Grappling with how ageing is embodied without reducing the body to a biomedical object; that ageing bodies are socially constituted.	a) Mask of ageing. Proposes that ambivalence towards ageing is less about socio-cultural influences and more about the internal sense of an 'ageless self' in a changing body b) Ageing and old age have intrinsic characteristics that should be understood and celebrated, e.g. existential/spiritual questions to be addressed at this time of life; time of maturation personally, physically, socio-culturally	Postmodern theories of fluid identities and lifestyles not constrained by normative social and biological categories; new technologies and roles to be celebrated; 'midlifestylers' into deep old age Embodiment theorising, the 'cultural' turn Jungian theory Gerotranscendence	Increasing homogeneity if 'anything goes' regardless of age. Lack of critique of consumerism and middle-aged values being imposed on later life. Reinforces Cartesian split between mind and body, e.g. 'not old' mind vs. 'old' body Gerotranscendence criticised as essentialist and attempting to measure and medicalise transcendence/spirituality
Ageing in place	*Ageing identity is located in place*			
	Relationships between characteristics of people and environments	Ageing-in-place a popular concept, but needs critical awareness of socio-political influences and consequences, such as 'care in the community' Ageing and place mutually constitutive	Functional, interactional analyses of person-environment fit Transactional connections between people and place Home as a 'process' not just a place; ageing identity actively negotiated in dynamic environments	Assessing individual person/environment fit too reductionist Superficial 'ageing-in-place' and 'care in the community' rhetoric can consign people to inadequate support and disabling environments

Table 12.2   Ideas associated with old age (adapted from Furstenburg 2002)

Negative ideas
- Mental, physical deterioration, and consequent changes
- General slowing and loss of energy or resilience
- Presence of chronic disease associated with ageing
- Deterioration of senses, hear less well, see less well, etc.
- Decline of cognitive function, consequent loss of activity (e.g., driving)
- Dependence on others
- Referring forward to death 'will have to face death soon'

Positive ideas
- Time to do things would like to do, have always wanted to do
- Time to focus on spirituality
- Ageing accompanied by internal development

Shifting the Debates: From Ageing as a Biomedical to a Social 'Problem'

Biomedical models are the dominant way of understanding ageing. Ageing is viewed as senescence at cellular and organism level, leading to a series of physical, psychological and social changes. Biomedical research on ageing tends to focus on diseases and disorders whose prevalence increases with age, from the 'young-old' Third Age into the frailty and senescence of the Fourth Age (around 85 years) (Baltes and Smith 2003).

Ageing is seen as an individual problem, and ageing of the population as a social problem. Dire warnings of increasing dependency ratios and sky-rocketing medical costs stem from this approach and feed general social views of ageing of populations as a potential crisis, including the colourful (but unwarranted) image of a 'demographic timebomb' that will overwhelm health and social services (Wilson 2000).

In social gerontology, various ways of understanding ageing have developed since the 1950s to theorise about how and why individual behaviour does (and should) change with ageing. Two early, contrasting theories were disengagement theory and activity theory.

Disengagement theory is the argument that ageing is associated with an inevitable and natural process of gradual withdrawal from work and people, and that this withdrawal is beneficial to 1) individuals because it suits their declining physical and psychological status, and 2) society because of the cyclical effects, e.g., retirement anticipates death (so is not disruptive to workforce) and creates room for new people in the workforce (Cumming and Henry 1961). Critics have viewed this theory as problematic because it explains old age in terms of its effects or death, and also because it condones indifference towards old age and isolation (Powell 2001).

Activity theory, in contrast, is the view that a *successful* old age can be achieved (and is defined) by engaging in and maintaining roles and relationships

(Neugarten, Havighurst and Tobin 1961). Proponents argue that ageing can be a lively and creative experience, and any roles or relationships lost should be replaced by new ones. They argue that disengagement is not a natural process, and it is an inherently ageist view to argue that it is (Katz 2000).

Critics point out that activity theory neglects issues of power, inequality, and conflict between age groups. Certain powerful interests in society find it useful to have age power relations organised in such a way that older people are expected to disengage (Powell 2001). Others contend that the problem is in the demarcation of activity being associated with success, so that any reduction of activity is seen as a kind of failure or problem.

Successful ageing continues to be the focus of more recent theories with strong biomedical roots, such as Rowe and Kahn's (1997: 439) model which proposes that 'successful' ageing can be distinguished from 'usual' ageing by three components:

1. Avoidance of disease and disability
2. Maintenance of high physical and cognitive function, and
3. Sustained engagement in social and productive activities

This model has generated much empirical work on 'staying healthy' into old age, and populist images of being able to choose between a wheelchair or cross-country-skiing at age 80, if you work hard enough at maintaining your health (Rowe and Kahn 1998). But along the way, the journey has become the destination, in that a few aspects of life that could underpin a good old age in all its diversity have instead become the definition and totality of 'success' (Holstein and Minkler 2003). These factors are also limited to the individual domain, taking little account of the social and environmental conditions across the lifespan that impact negatively on health, including the places (and 'place' in society) of ageing (McHugh 2003).

Applying narrow, fixed criteria of 'success' means that many people will 'fail' at ageing, yet in the few studies that ask older people about their experience of 'successful ageing', the majority claim success even in the presence of illness (e.g., 75 per cent of a UK population sample felt successful where the model would find only around a fifth measuring up – Bowling 2006).

The model of 'selective optimisation with compensation (SOC)' emerging out of the Berlin Study on Ageing (Baltes and Baltes 1990), highlights adaptive strategies that individuals can use to successfully manage the changes and losses of ageing. For example, as a person becomes frail, the quantity of their social contacts may decrease, but the quality can be maintained by *optimising* strategies, such as maintaining reciprocal emotional support exchanges, *selecting* out less important social connections, and getting instrumental support from paid help to *compensate* for loss of function (Lang and Carstensen 1994). Strategies such as moving into retirement villages or 'age-appropriate' housing (one-level, low-maintenance) can be viewed as a geographical sort of adaptive SOC, also providing refuge from the pervasive ageism by being in a place where 'everybody is old so nobody is old' (McHugh 2003: 181). They can also be critiqued as forming a geography of wealthy consumers keeping the ageing and declining body hidden in specialist places at the margins of society (Tulle-Winton 1999).

What Do Older People Say?

Occasional studies of older people's views of 'successful ageing' (e.g., Bowling 2006, Duay and Bryan 2006) find multi-dimensional criteria of 'success', ranging from financial security to spirituality, abilities to cope with loss or maximise opportunities to learn, appreciate aspects of home and neighbourhood, and engage in creativity (whether in the art-studio, garden or kitchen), a rich complexity at odds with the narrow ideals of 'success' as an absence of disability.

There is a similarly wide range of aspects of ageing noted in our sense of place research with older people in response to the question, 'What is good about growing older?' The negative responses of around a quarter of them are noted above, in terms of the ambivalence people therefore have around the process of ageing. The largest group (43 per cent), however, were able to think of good things, and a further third gave a mixed account of good and bad or a process of acceptance and adaptation. Good things about growing older were rather at odds with the emphasis on health in the successful ageing framework. Only two mentioned physical or cognitive health. 'Engagement in social and productive activities', the third criteria of successful ageing, was rarely mentioned; one person liked 'having things to do' but also appreciated 'nice peaceful times like today'. Two liked opportunities to learn new things, but less because that was productive than because it defied age stereotypes: 'Can still learn even when you're 80'.

Far more of our participants valued *not* being 'productive', savouring 'not having to go to work on a cold wet day', 'lying in bed of a morning if you feel like it', the choice to 'do things in your own time or put it off altogether' and the pleasure of 'knowing I…don't have to join anything'.

It is worth noting that our interviewer asked the 'growing older' question at the end of our project, which had involved visiting participants up to eight times over a year. This gave a level of rapport that may have facilitated responses beyond the formulaic, such as the less 'socially desirable' responses like lying in bed, or putting things off. A one-off interview or tick-box questionnaire may be less likely to elicit such responses.

Most good things about growing older for our participants related to feelings and attitudes. Several valued their increased acceptance and tolerance of themselves and others, as this 84-year-old woman says, 'I guess you don't have the pressure, such as getting ahead. More accepting of people and their ways, more amenable to people'. Participants enjoyed a sense of freedom and independence with age, being 'more free to do what you want'. One man talked about surviving tough life experiences which had made for a greater sense of independence, and two mentioned the financial freedom of 'no mortgage' and feeling 'privileged'. One 83-year-old woman enjoyed freedom from the social pressure she used to feel 'to have my hair dyed or curled'. Wisdom, laughter, peace of mind and faith were key feelings or attributes many participants mentioned. One 77-year-old woman was pleased that she felt she was (at last?) 'growing up and managing myself.'

For the group that gave a mixed account of 'good' and 'bad' in response to the 'growing older' question, there were many 'both/and' responses, acknowledging both benefits and costs, as this 82-year-old man said, 'Well, I suppose it's got its ups and downs. Take life as it comes – it doesn't worry me. Wouldn't say it's good to get old, but wouldn't want to stay young.' On the negative side, some mentioned death, health and being able to 'do less'. On the positive side, attitude was again important – 'not worrying about the future' or having a 'good frame of mind', as was the feeling that 'People expect less of you when you're older'. Within many of the responses of the 'mixed' group was an attitude of acceptance and inevitability about growing older, saying you just have to 'take it as it comes', and recommending you 'don't think about it'. An 80-year-old man represents the typically rich mixture of both positive and negative responses to growing older: 'I can't think of too much positive. A big psychological step turning 80, had a great party, but my brothers died in 80s so I think life is coming to an end. With age, you get more generous, more understanding of others, accept their failings more readily.'

Tensions between Fighting against Inequality and Celebrating Diversity

There is tension between fighting to have social differences recognised and valued without stigmatising people and setting them apart. These battles are ongoing within areas such as the disability sector, and increasingly, within theorising about age as a social difference.

Political economists emphasise old age as a product of social structural factors whereby it is *society* that creates problems of old age rather than physiology (Estes 2001), in line with the arguments of disability theorists that it is society, rather than individual physiology, that 'disables' people. For example, the idea of 'structured dependency' (Townsend 1981), suggests reliance on welfare benefits and pensions amongst older people and those with disabilities is created by excluding them from economically productive work, through lack of workplace flexibility to accommodate diversity, compulsory retirement, and other effects of ageism and ableism (Twigg 2004).

Feminist gerontologists have examined old age in relation to other societal constructs to understand how power imbalances shape the theoretical construction of old age. Arber and Ginn (1995) point to a double jeopardy of gendered ageing and sexual discrimination. Whereas old age in men is associated with power, wealth and social (though not physical) dominance, old age in women has especially strong negative connotations. For women sexual attractiveness is seen to reside in youth: 'The older female body is both invisible – in that it is no longer seen – and hypervisible – in that it is all that is seen' (Twigg 2004: 61).

Foucauldian gerontology attempts to understand old age through conceptual exploration of power and knowledge (Katz 1996), how surveillance practices marginalise, normalise and shape experience. Sociologist Janet Heaton (1999: 773) offers a Foucauldian analysis of the discourse of 'informal care' within ageing and disability, in which ordinary social contacts have been turned into 'carers' as 'care in the community' has been transformed into 'care by the community'. She sees a devolution of the medical 'gaze', in that informal carers are responsible for 'overseeing' the health of those they care for, while being overseen themselves by the formal care systems of the state (such as medics and social workers). Similarly, Biggs and Powell (2001) use Foucauldian frameworks to explore ageing identity formation, social policy narratives and professional practices.

What Do Older People Say?

The issue of transport is an example of the way societies structurally 'disable' older people. In our research on sense of place, participants spoke about barriers to maintaining their 'social spaces' – the social worlds of which they had been a part. The most common way their social networks had been reduced was through the deaths of friends and family, followed by their own or others' health problems. But driving was the third most common reason given for decreasing social spaces, with people driving less, not driving at night, or not driving at all. It could be argued that driving less should have no impact on social spaces if adequate, affordable public transport suitable to wide-ranging levels of ability was available, as this 81-year-old woman points out: 'I don't drive now, buses not very frequent and they moved the bus stop, spine pain harder to walk around the block [to the new bus stop].' This meant both that she stopped doing things and also felt a loss of independence, 'It's hard while I can't drive, dependent on people – don't want to be a nuisance.'

How Language Shapes and is Shaped by Ageing

Language is constitutive of social reality, according to social constructionist and discursive theorists. The negative associations of 'oldness' and 'disability' are evident in the ambivalence about embracing these words in relation to one's identity. Even what words should be used to denote older people are contested. Bytheway (2000) thinks the word 'elderly' is superfluous and should not be used, suggesting it is even as offensive as use of the term 'moron' in disability studies. Others see 'elderly' as respectful, a term that older people use to talk about themselves, and Gibson (2000) thinks the word should be reinvigorated. The phrase 'older people' is widely used in social and health services, and signals a relative state, rather than the absolutism of a contested word like 'aged'. Andrews (2000) views *agefulness* as something to be proud of and calls for the reinvigoration and reinvention of such language.

How does socio-cultural stigma become everyday talk? One approach is to consider personal experiences of *socially and culturally transmitted ideas* about ageing. For example, Furstenburg (2002) proposes that the high level of social agreement about negative expectations for growing older suggests such 'models' of ageing are culturally transmitted. Observations of others and other sources of information contribute to both generalised trajectories of ageing (as it happens to others) and to the drawing of implications more specifically for the self. This helps to explain why people do not 'feel' themselves to be old, but nevertheless have goals for future behaviour in old age such as financial planning, health behaviour, decisions about living near relatives or other forms of support, or renovations to their houses. The focus is not just on what people *do*, but what they *imagine* in conjunction with what they are doing as these personal and general models are continually revised in an active, interpretive process. This accounts for the difference between changes foreseen for older people in general, and those expected for oneself (Furstenberg 2002).

What Do Older People Say?

There has been some theorising around the role of *discourse* in framing age identity, stages of life, and the meaning of old age (Nikander 2000). This is particularly related to the way people talk about old age and the process of growing older. For example, Jolanki et al. (2000) identify their 90+-year-old subjects using two contrasting repertoires or commonly understood clusters of resources for describing age: old age as choice and as necessity. These repertoires are selectively drawn upon and reworked according to the setting. Within the 'necessity' repertoire, it was 'self-evident' that the essence of old age is deterioration, whereas the contrasting 'choice' repertoire was built up by the same older people to undermine the 'necessity' view, suggesting more varied and positive definitions of old age that one can choose. Jolanki et al. (2000) locate this contrast as being within individuals' minds, and not just between cultural views and individual experiences, and emphasise old age as a dilemma. They suggest this to be an opportunity, in that considering the dilemmatic nature of old age 'frees us from the hopeless task of trying to find the "right" definition of old age' (Jolanki et al. 2000: 371).

Nikander (2000: 351-2) argues that age identity 'can be viewed as a *discursively achieved process*' whereby talk is a way in which people manage and solve conflicting 'age identities'. Because language constructs, limits, and guides our understanding of our worlds and ourselves, it is actively used in everyday meaning construction. Similarly, Jones (2006) suggests treating old age as 'socially constructed in discourse' and uses positioning theory to examine how people position themselves as 'old' or 'not old' according to the business of the conversation or interaction at hand. This positioning is variable, multiple and active. When an older person does not use the 'older' identity, it is not an act of foolish denial (as an exasperated health service worker might claim), but is a positioning in a particular conversation with associated rights and duties (e.g., If I'm not 'old', I don't have to attend the boring, aged care activity programme that you're trying to get me to go to). This also helps explain why people talk about the individual and social *experience* of growing older in such contradictory ways. We could thus draw on Judith Butler (1990), to think of older age as a kind of performativity whereby people draw on a range of often contradictory, tacit, collective and individual ideas about age to frame, constrain and enable their 'performance' of old age in ways that are never quite perfectly reproduced (Butler 1993).

Bringing Back the Body

We have discussed four areas of overlap between scholarship on ageing and theorising disability, namely (1) ambivalence regarding stigmatised identities, (2) the trajectory of the debates across biomedical and social science disciplines, (3) the issues of structural inequality and inequity, and (4) theorising how language is constitutive of these 'problems'. It is possible to draw much of this together in highlighting some final key concerns. There is a need to embrace the play of tensions between the individual and social worlds, the material and the discursive, between time and place. The body, including its pain and impairment, must not be excluded from socio-political analyses of ageing and disability, and we need to put embodied ageing and disability in its place – within specific environments and geographies.

With the arguments around the social, cultural, and societal construction of old age and against reductionist physiological approaches to ageing, the presence of bodies in social gerontology has been uncertain and muted. Twigg (2004) points out that a political-economy approach which emphasises old age as a product of social structural factors is at risk of avoiding the bodily territory of deterioration and decline, effectively handing it over to the biomedical lobby. The subject of the ageing body has been taken up to some extent by feminist gerontologists, as women researchers have found themselves to be increasingly grounded in age oppression and struggled against the reduction of women to mere bodily characteristics, but Twigg (2004) argues that the literature on the ageing body is generally focused on late middle years (50s to 70s) rather than 'deep old age.' She sees parallels with gender and disability studies where there has been 'a similar recovery of the ways the body is socially constituted, destabilising earlier conceptual distinctions between sex and gender, impairment and disability' (Twigg 2004: 61). The cultural turn, she suggests, has led to a refocus on the body which derives from earlier social constructionism but more radically shows how the body itself is socially constituted.

Where social gerontologists *have* addressed the body in old age, a vigorous debate has developed between two approaches: a) the idea of the 'mask' of ageing or tension between the (essential) self and the physical material aspects of ageing (the body), and b) the concept of intrinsic ageing, the idea that there are discernible differences between older age and other parts of the life course, so that old age has particular characteristics, activities and experiences.

The Mask of Ageing

On one side is the argument that there is very little that is intrinsic to later life. Postmodern gerontologists have argued that we could use the image of the 'mask' of old age covering the essential identity of the person underneath to alert ourselves to the tension between the external appearance of the body, face, and functional capacities, and the internal or subjective sense of experience of personal identity (Featherstone and Hepworth 1989). Old age provokes a lack of fit between what the observer sees and the experience of self; at the extreme, the youthful inner self trapped in an older body. This mask of old age is also social and symbolic; older people are often lumped into a group as if they were homogeneous and 'fixed' to roles without resources. The aged body (and general perceptions of older bodies) becomes a barrier that prohibits engagement of the youthful self with consumer society and multiple lifestyles.

A postmodern perspective would instead see age as fluid with possibilities, not constrained by medical models of decline, with many pointing to the use of technologies in the post-modern era to modify the experience or appearance of ageing (Powell and Longino 2001). Now even lifecourse categories are becoming increasingly blurred, and some would argue that those in an old age stage of life do not necessarily have distinctive qualities from other stages because the trends of postmodernity (e.g., 'gray' lifestyle consumerism, 'active' retirement, technological advances, and the possibility of virtual identities) have multiplied the options open to older people. This means that people are no longer bound by biological and social constraints but can maintain 'midlifestyles' as long as possible, even into deep old age (Featherstone and Hepworth 1983). Biggs (2005) notes the irony here that while there are more diverse forms of old age, there is greater homogeneity/uniformity between different age groups as one ceases to distinguish from another.

Kaufman's (1986) idea of the 'ageless self' emphasises the idea of continuity of self regardless of one's passage through the stages of life, distinguishing between the ageing body and the self. Kaufman argues that old people are aware of popular conceptions of old age and how these relate to appearance, but choose to reassert and find meaning in the idea of a continuing 'un-ageing' self. She would argue that older people rarely see themselves as old because the inner sense of self is ageless in spite of physical and social changes associated with ageing.

Intrinsic Ageing

Andrews (2000) challenges the idea of a continuity of identity, arguing that this is a form of agelessness (and therefore ageist). She sees this as a form of Cartesian dualism which has come under attack in many areas of study but not in old age. While agreeing that most older people 'don't feel old', she sees this 'ageless self in an ageing body' argument as an attempt to deny processes of adult ageing, which are erroneously formulated around the experience of illness.

An alternative approach is the argument that there are discernible differences between later life and other parts of the life course. Opinion is varied on whether these intrinsic differences are primarily social, or from processes of maturation as life progresses. For example, Jungian psychology identifies specific challenges and existential questions associated with the later part of the life course, so that those who cling to one stage of life while entering another will suffer psychological dysfunction (Biggs 2005).

Spiritual and existential aspects of later life are the focus of theorising such as 'gerotranscendence' (Tornstam 1997), the idea that there is a fundamental alteration of consciousness in old age leading to greater wisdom and maturity and a break with the materialist and rational views of mid life. Attempts to theorise how this process of reorganisation and reconstruction might occur have been criticised as too essentialist and normative, an attempt to medicalise mysticism (Jönson and Magnusson 2000).

American philosopher David Norton (1976) argues that while adult ageing consists of stages it is *experienced* as uniform, and in retrospect it is difficult to see anything other than continuity, because preceding ages are understood through the lens of the stage one is currently in. Thus older people may become aware that they have a new set of priorities that distinguish them from other groups, yet perceive themselves as relatively unchanged. Old age is thus not deterioration but 'rebirth under novel criteria' (Norton 1976: 165).

Cole's (1983) historical analysis explores the move over time from a view of ageing that held in creative tension the ambiguity, contingency, intractability and unmanageability of human life, towards now more polarised views of old age as a 'problem' (associated with physical and mental deterioration, dependency, poor health, frailty) or imbued with positive mythology (that older people are, or should be, healthy, sexually active, engaged, productive and self reliant). Perhaps it is time to recover a more complex, embodied view of ageing in all its nuanced ambiguity.

Embodied Ageing in Place

A key tenet of our argument is that embodied ageing and disability are best understood *in place*, within specific social and physical environments and geographies. Theorising on the relationships between people and environments is a growing area of ageing research. Widespread policies supporting 'ageing in place', where people remain in their homes and communities for as long as possible, are favoured by policy-makers and health providers and by many older people themselves (Pastalan 1990). Typically, ageing in place is linked to maintaining independence, and the appeal of a less costly option than institutional care. However, critical analysis of the move to ageing in place and 'care in the community' (a phrase used in both the disability and eldercare sectors) highlights the mix of rhetoric that valorises 'community' or 'family' connection and support, independence and choice, on the one hand, with cost-savings of deinstitutionalisation and a lack of real choice on the other (Heaton 1999, Wilson 2000).

While claims that people prefer to 'age in place' abound (Frank 2002), the term is ambiguous. It is a complex process, not merely about attachment to a particular home or neighbourhood, but where the older person is continually reintegrating with places and renegotiating meanings and identity in the face of dynamic landscapes of social, political, cultural and personal change (Andrews et al. 2007).

Relationships between ageing and place have been explored in functional analyses of person-environment fit (Lawton 1982), which assess the interactions between a person's relative 'competence', in terms of mobility for example, and the characteristics of the environment, for example in terms of accessibility. Interactions between these functions are reciprocal, in that the less 'competent' the person, the greater the influence of the environment on their behaviour, whereas a more competent person may be better able to utilise the resources of any environment in the service of their personal needs (Lawton 1999). Similarly, geographers have shown how a well-resourced environment, including appropriate physical design, can enhance the physical independence of an individual.

We need to develop our understanding of the spaces in and through which the 'difference' of old age is produced, embodied, and experienced. Places are shaped by the intimate relations between people, but also shape them. For example, moving the care of people to the more private, less visible, and often more isolated spaces of private dwellings has implications for the experiences of both caregivers and care-receivers (Andrews et al. 2007). Homes are not always tranquil havens, but can be sites of conflict. Care at home can intensify complex interrelations of power and family relationships, often in larger contexts of inadequate alternative care provision, as the growing literature on elder abuse attests (Lowenstein 2009). As Andrews et al. (2007: 12) point out: 'Too frequently, there is a tendency to treat "place" simply as a context (clinical or living), rather than seeing it as productive of particular outcomes for older adults, as well as being shaped by them.'

Places are physical, but also operate on social and symbolic levels in interconnected ways. A more 'transactional' analysis between people and places builds on Graham Rowles' geographic exploration of the meaning, security, and sense of identity, and thus well-being, that older people draw from their homes (Rowles 1993, Rowles and Ohta 1983). A person's 'home' is a physical location, but is also imbued with wide-ranging and ever-changing social and symbolic meanings, to the extent that talk of 'home' is a constant process and negotiation of meanings (Wiles 2003, 2005a), incorporating not just a physical house but also its setting (Peace et al. 2006).

Places are also connected to other places at all sorts of levels. Policy decisions on health or social services made at a national level can directly impact on what happens in a person's home, in terms of whether disability or frailty can be well-supported or autonomy maintained (Wiles 2005b). Power relations are both expressed through places and shape those places, with older people often not having a powerful voice in neighbourhood and infrastructure planning debates. Geographer Glenda Laws (1993: 672) describes the urban built environment in the US as both a 'cause and an effect of ageist attitudes' with some urban areas inhospitable to older people (and people with disabilities), and others enacting segregation of age-groups (in retirement villages far from school-yards or playgrounds). Initiatives such as the WHO's Global Age-Friendly City project, and growing emphasis on positive ageing and creating healthier, more accessible environments for older people may change this. Though it remains to be seen how well the voices of older people themselves are incorporated into these, many of those aspects of physical and social environments that would be more inclusive and amenable for older people would also impact positively on people living with disabilities.

Concluding Thoughts

Old age is spoken of with an ambivalence and stigma that those in the disability field can no doubt relate to. This chapter has explored some of the sources of that ambivalence and the way the identities of 'old' and 'not old' are theorised and enacted. Biomedical models remain dominant in the 'treatment' of both ageing and disability, but critical social theorists continue to challenge and problematise that dominance, particularly as it informs wide-ranging social policy and practice. There is also some turning to older people themselves to enrich our nuanced understandings of later life (though not yet to sufficiently let 'them' lead research and policy agendas). The contested identities of age heighten the tensions between fighting against inequality and celebrating diversity that are also evident in the disability sector. The language we use and how socio-cultural ideas become personal also echo the discursive understandings of disability.

Reorienting the debates on ageing towards bodies and places seems a way to draw together and extend existing scholarship. How can the socially constituted body experience frailty and pain without disappearing back into the medical domain? How can the dynamic processes by which spaces and places shape and are shaped by ageing populations be theorised and influenced? How might disability and ageing be similar and different in these processes? Together, we need to develop our understanding of the spaces in and through which the 'differences' of old age and disability are produced, embodied, and experienced, so that the identities of agefulness and disability can cease to be spurned.

References

Andrews, G.J., Cutchin, M., McCracken, K., Phillips, D.R. and Wiles, J.L. 2007. Geographical gerontology: the constitution of a discipline. *Social Science and Medicine*, 65(1), 151-168.
Andrews, M. 2000. Ageful and proud. *Ageing and Society*, 20(6), 791-795.

Arber, S. and Ginn, J. 1995. *Connecting Gender and Ageing: A Sociological Approach.* Buckingham: Open University Press.

Baltes, P.B. and Baltes, M.M. 1990. Psychological perspectives on successful aging: the model of selective optimization with compensation, in *Successful Aging: Perspectives from the Behavioral Sciences,* edited by P.B. Baltes and M.M. Baltes. Cambridge: Cambridge University Press, 1-34.

Baltes, P.B. and Smith, J. 2003. New frontiers in the future of aging: from successful aging of the young old to the dilemmas of the fourth age. *Gerontology,* 49, 123-135.

Biggs, S. 2005. Beyond appearances: perspectives on identity in later life and some implications for method. *The Journals of Gerontology Series B: Psychological Sciences and Social Sciences,* 60, S118-S128.

Biggs, S. and Powell, J.L. 2001. A Foucauldian analysis of old age and the power of social welfare. *Journal of Aging and Social Policy,* 12(2), 93-111.

Bowling, A. 2006. Lay perceptions of successful ageing: findings from a national survey of middle aged and older adults in Britain. *European Journal of Ageing,* 3, 123-136.

Butler, J. 1990. *Gender Trouble: Feminism and the Subversion of Identity.* New York: Routledge.

Butler, J. 1993. *Bodies that Matter: On the Discursive Limits of 'Sex'.* New York: Routledge.

Bytheway, B. 2000. Youthfulness and agelessness: a comment. *Ageing and Society,* 20(6), 781-789.

Cole, T.R. 1983. The 'enlightened' view of aging: Victorian morality in a new key. *The Hastings Centre Report,* 13(3), 34-40.

Crooks, V.A., Chouinard, V. and Wilton, R. 2008. Understanding, embracing, rejecting: women's negotiations of disability constructions and categorizations after becoming chronically ill. *Social Science and Medicine,* 67(11), 1837-1846.

Cumming, E. and Henry, W.E. 1961. *Growing Old: The Process of Disengagement.* New York: Basic Books.

De Beauvoir, S. 1970. *Old Age (La Vieillesse)* P. O'Brian, Trans. London: Weidenfeld and Nicolson.

Duay, D.L. and Bryan, V.C. 2006. Senior adults' perceptions of successful aging. *Educational Gerontology,* 32, 423-445.

Estes, C.L. 2001. *Social Policy and Aging: A Critical Perspective.* Thousand Oaks: Sage.

Featherstone, M. and Hepworth, M. 1983. The midlifestyle of 'George' and 'Lynne': comments on a popular strip. *Theory, Culture and Society,* 1(3), 85-92.

Featherstone, M. and Hepworth, M. 1989. Ageing and old age: reflections on the postmodern life course, in *Becoming and Being Old: Sociological Approaches to Later Life,* edited by B. Bytheway et al. London: Sage, 143-157.

Frank, J.B. 2002. *The Paradox of Aging in Place in Assisted Living.* London: Bergin and Garvey.

Furstenburg, A.-L. 2002. Trajectories of aging: imagined pathways in later life. *The International Journal of Aging and Human Development,* 55(1), 1-24.

Gibson, H.B. 2000. It keeps us young. *Ageing and Society*, 20(6), 773-779.

Heaton, J. 1999. The gaze and visibility of the carer: a Foucauldian analysis of the discourse of informal care. *Sociology of Health and Illness*, 21(6), 759-777.

Holstein, M.B. and Minkler, M. 2003. Self, society, and the 'new gerontology'. *The Gerontologist*, 43(6), 787-796.

Hurd, L.C. 1999. 'We're not old!': Older women's negotiation of aging and oldness. *Journal of Aging Studies*, 13(4), 419-439.

Jolanki, O., Jylhä, M. and Hervonen, A. 2000. Old age as a choice and as a necessity: two interpretive repertoires. *Journal of Aging Studies*, 14(4), 359-372.

Jones, R. 2006. 'Older people' talking as if they were not older people: positioning theory as an explanation. *Journal of Aging Studies*, 20(1), 79-91.

Jönson, H. and Magnusson, J.A. 2001. A new age of old age? Gerotranscendence and the re-enchantment of aging. *Journal of Aging Studies*, 15(4), 317-331.

Katz, S. 1996. *Disciplining old age: The Formation of Gerontological Knowledge*. Charlottesville: University Press of Virginia.

Katz, S. 2000. Busy bodies: activity, aging and the management of everyday life. *Journal of Aging Studies*, 14(2), 135-152.

Kaufman, S.R. 1986. *The Ageless Self: Sources of Meaning in Late Life*. Madison: University of Wisconsin Press.

Kite, M.E., Stockdale, G.D., Whitley Jr, B.E. and Johnson, B.T. 2005. Attitudes toward younger and older adults: an updated meta-analytic review. *Journal of Social Issues*, 61(2), 241-266.

Lang, F.R. and Carstensen, L.L. 1994. Close emotional relationships in late life: further support for proactive aging in the social domain. *Psychology and Aging*, 9(2), 315-324.

Laws, G. 1993. 'The land of old age': society's changing attitudes toward urban built environments for elderly people. *Annals, Association of American Geographers*, 83, 672-693.

Lawton, M. 1982. Competence, environmental press, and the adaptation of older people, in *Aging and the Environment: Theoretical Approaches* edited by M. Lawton, P. Windley and T. Byerts. New York: Springer, 33-59.

Lawton, M. 1999. Environmental taxonomy: generalizations from research with older adults, in *Measuring Environment across the Life Span: Emerging Methods and Concepts* edited by R.J. Scheidt and P.G. Windley. Washington: American Psychological Association, 91-124.

Lowenstein, A. 2009. Elder abuse and neglect – 'Old phenomenon': new directions for research, legislation, and service developments. *Journal of Elder Abuse and Neglect*, 21(3), 278-287.

McHugh, K.E. 2003. Three faces of ageism: society, image and place. *Ageing and Society*, 23, 165-185.

Minichiello, V., Browne, J. and Kendig, H. 2000. Perceptions and consequences of ageism: views of older people. *Ageing and Society*, 20, 253-279.

Neugarten, B.L., Havighurst, R.J. and Tobin, S.S. 1961. The measurement of life satisfaction. *Journal of Gerontology*, 16, 134-143.

Nikander, P. 2000. 'Old' versus 'Little Girl': a discursive approach to age categorization and morality. *Journal of Aging Studies*, 14(4), 335-358.

Norton, D.L. 1976. *Personal Destinies: A Philosophy of Ethical Individualism.* Princeton: Princeton University Press.

Oldman, C. 2002. Later life and the social model of disability: a comfortable partnership? *Ageing and Society*, 22, 791-806.

Pastalan, L.A. 1990. *Aging in Place: The Role of Housing and Social Supports.* New York: Haworth Press.

Peace, S.M., Holland, C. and Kellaher, L. 2006. *Environment and Identity in Later Life.* New York: Open University Press.

Powell, J.L. 2001. Theorising social gerontology: the case of social philosophies of age. *The Internet Journal of Internal Medicine*, 2(1).

Powell, J.L. and Longino, C.F. 2001. Towards the postmodernization of aging: the body and social theory. *Journal of Aging and Identity*, 6(4), 199-207.

Rowe, J.W. and Kahn, R.L. 1997. Successful aging. *The Gerontologist*, 37(4), 433-440.

Rowe, J.W. and Kahn, R.L. 1998. *Successful aging.* New York: Dell.

Rowles, G.D. 1993. Evolving images of place in aging and 'aging in place'. *Generations*, 17(2), 65-70.

Rowles, G.D. and Ohta, R.J. 1983. *Aging and Milieu: Environmental Perspectives on Growing Old.* New York: Academic Press.

Tornstam, L. 1997. Gerotranscendence: the contemplative dimension of aging. *Journal of Aging Studies*, 11(2), 143-154.

Townsend, P. 1981. The structured dependency of the elderly: a creation of social policy in the twentieth century. *Ageing and Society*, 1(1), 5-28.

Tulle-Winton, E. 1999. Growing old and resistance: towards a new cultural economy of old age? *Ageing and Society*, 19, 281-299.

Twigg, J. 2004. The body, gender, and age: feminist insights in social gerontology. *Journal of Aging Studies*, 18(1), 59-73.

Wendell, S. 1996. *The Rejected Body: Feminist Philosophical Reflections on Disability.* New York: Routledge.

Wiles, J.L. 2003. Daily geographies of caregivers: mobility, routine, scale. *Social Science and Medicine*, 57(7), 1307-1325.

Wiles, J.L. 2005a. Conceptualising place in the care of older people: the contributions of geographical gerontology. *International Journal of Older People Nursing*, 14(8b), 100-108.

Wiles, J.L. 2005b. Home as a new site of care provision and consumption, in *Ageing in Place* edited by G.J. Andrews and D.R. Phillips. London: Routledge, 79-97.

Wiles, J.L., Allen, R.E.S., Palmer, A.J., Hayman, K.J., Keeling, S. and Kerse, N. 2009. Older people and their social spaces: a study of well-being and attachment to place in Aotearoa New Zealand. *Social Science and Medicine*, 68(4), 664-671.

Wilson, G. 2000. *Understanding Old Age: Critical and Global Perspectives.* London: Sage.

Chapter 13

Biometric Geographies, Mobility and Disability: Biologies of Culpability and the Biologised Spaces of (Post)modernity

Joanne Maddern and Emma Stewart

Introduction

The widespread changes occurring in the way mobility is organised globally means that geographers researching disability are increasingly concerned with the relationship between disabled persons and technologies as well as questions of citizenship and social inclusion (Crooks, Dorn and Wilton 2008). Research on the spatialities of disability has focused upon the ways in which disabled people's lives are constitutive of interactions between their (perceived) impaired bodies and the places those bodies inhabit and move through (Imrie 2001). This has included a focus upon the local landscape to explore the physical construction and demarcation of urban spaces as well as elucidating ableist values etched across urban and rural landscapes.

Recognising the need to advance beyond the wider built environment, and an examination of mobility constraints at the local level, this chapter extends the analysis to explore developments in the regulation of cross-border movements. The primary focus is upon the development and implementation of biometric technologies such as iris scanning, DNA matching, facial recognition technology and fingerprint capture that are being employed to control and monitor global mobility. The chapter explores these advances in biometric technologies as a means of unearthing potential implications for disabled groups. Through references to a small, qualitative research project carried out in 2005, it aims to give empirical examples of ableist ideas, practices and institutions as they find concrete and discursive form within the biometrics industry.

Biometric identifiers are based upon physiological patterns such as fingerprints, iris, face or voice recognition or behavioural patterns such as hand-written signature verification. The focus is upon the physiological or unique biological make-up of an individual which is used as a means of authentication or verification of identity. The process involves the collection of raw biometric data (e.g., fingerprint or iris scan) which is then processed into a template and is stored in the memory of a biometric device or a centralised database. There has been growing policy and academic interest in the development and implementation of biometric technologies such as iris scanning, DNA matching, facial recognition technology and fingerprint capture in recent years (Agre 2001), with many different practical applications for this technology being suggested (for a comprehensive discussion see Thomas 2005).

The rapid diffusion of biometric technologies has been the result of the interaction between policy actors in the US and the European Union, in an international network of post 9/11 activities (Petermann et al. 2006). Countries within Europe are committed to introducing biometric passports. These include two mandatory biometric identifiers, facial recognition and fingerprints, with iris scans possibly introduced voluntarily at a later stage. Currently a facial image, a digitised copy of the standardised passport photo, is set as a minimum standard by the International Civil Aviation Organisation (ICAO). A facial scan, however, is different. These are taken in controlled environments where a camera takes a picture of the person recording up to 1,840 distinctive features of their face. Such advanced technology is promoted because 'biometric characteristics cannot be faked, lost or stolen – or so it is claimed; hence the belief that biometrics can provide foolproof, positive identification' (Van der Ploeg 2003: 86).

A UK governmental programme has been implemented at a number of national and regional scales to assess the potential effectiveness of such systems for future use on the general population. This has included plans for the inclusion of biometric data on passports and other 'identity' documents such as National ID cards. Consultation on this issue has led to the consideration of the potential cost of the ID card scheme, pilot schemes (such as the UK's Iris Recognition Immigration System (IRIS) which was trialled at Heathrow in 2002) as well as prompting significant resistance towards the plans on the basis of eroding basic privacy rights (Bevan 2008; BBC News 2008). There are also plans for the targeted use of biometric systems for use on specific social groups such as asylum seekers as set out in section 126 of the Nationality, Immigration and Asylum Act 2002.

There has been significant academic research on biometric technologies from a technological perspective (Ball 2002) and much work carried out on more general technologies of surveillance at airports and other key spaces of mobility (Adey 2004a, 2004b), but little on biometrics and disability. The attractiveness of biometric technologies to authorities is due to 'their quality as a unique, permanent and universal imprint of a person's identity' (Thomas 2005: 2). Based upon an existing template, a person presenting themselves is recognised by the system and they are regarded as 'known'. This is particularly appealing to immigration authorities when dealing with asylum applicants because 'if a person shows up with nothing with them but the clothes they wear and the story they offer, it would, of course, be a golden solution to be able to produce from that person's body an identity that is independent of that story, and yet undeniably belonging to that person' (Van der Ploeg 1999: 300).

The body is regarded as a text which can be read to ascertain the identity of an individual. Biometrics is thus based upon the notion that identity is fixed and pre-given so determining the identity of an asylum seeker or migrant is the process of verifying whether their 'claimed' identity matches their 'true' identity. And so 'coupled with the idea of travellers' infallible identification based on their bodily features, this calculus, and the identities it confers on a person, comes to enjoy an aura of legitimacy and objectivity' (Häkli 2007: 140). However, as we will argue in this chapter, people with disabilities have not been given as much prominence in biometric trials, along with other 'minority' groups such as asylum seekers. In this sense, a two-tier system is unwittingly being created – the legible and mobile and the illegible and therefore immobile. This will, of course, exacerbate the mobility restrictions already faced by some individuals with disabilities.

National restrictive policies and biometric technologies are thus powerful tools to prevent the entry of unwanted populations and control the movement of certain bodies within national borders. But whilst biometric systems have been heralded as a panacea to the security problems currently faced by Western governments, this chapter argues that they are complicit in a resurgence in biologism whereby 'biologies of culpability' are increasingly marginalised from the privileged spaces of (post)modernity through a newly legitimated embodied exceptionalism. Despite attempts by producers to emphasise the effectiveness and accuracy of the emerging biometrics market, recent critiques have suggested that software-sorting systems are increasingly complicit in new national and global geographies of inequality (Rose 2003, Graham 2005, Amoore 2006).

Indeed, there are several reasons why disability researchers should be concerned about the growing rise in biometric technologies. Biometric technologies are powerful techniques implemented to maintain national borders and territory. Nonetheless, 'territorialisation is often subtle, including signs and symbols or socio-cultural codes which are, potentially, powerful demarcaters of difference' (Imrie 2001: 233). It has been noted that surveillance 'operates as a medium to (de)classify, separate, organise, control and regulate life' (Weibe 2008: 337). A report which investigated EU countries' experiences of ID card systems found that immigration officials and police officers checked the identities of people from ethnic minorities disproportionately (Beck and Broadhurst 1998). Questions therefore arise as to the potential exclusionary and discriminatory implications of new forms of surveillance like biometric technologies upon marginalised groups, including disabled persons. Previous disability research has highlighted the ways in which the body is often neutered and neutral with sex, gender, race or physical difference ignored (Imrie, 2003). Given the primary focus of biometric technologies upon the body, developments in this area should be of particular interest and concern to disability researchers.

Many questions about the nature and effects of biometric systems upon disabled persons remain largely unanswered at the empirical level by academics and little work has been carried out to test the rather grand theoretical claims made. Accordingly, this chapter draws on a selection of interviews with consultants, producers and users of biometric technologies in the United Kingdom carried out in 2005. The research project investigated the relationships between the body, migration and identity through a specific focus on biometric technologies. This chapter critiques the unfettered use of biometric technologies without human intervention by focusing upon the operationalisation and limitations of the technologies. The discussion explores the application of biometric technologies to 'vulnerable' groups such as disabled individuals and asylum seekers. A key theme that emerges from an analysis of the interviews is the fallibility of biometric systems, particularly for disabled users as well as any societal group that deviates from the 'norm'. By 'norm' here, we are referring to a state of being in which significant deviations from the median qualities of the population are not evident – what we might refer to as Da Vinci's *Vitruvian Man* in which all proportions and faculties correspond to some predefined quantitative measure of 'normality'. Despite the now dated notion evident in Da Vinci's ideas of 'perfection', the trend for biometrics, whilst not explicit, contains the potential for a new biologism to emerge in which anthropometry is replaced by a new notion that there is a universal set of proportions for the human body. Rather than the body being seen as a microcosm of the universe, as Da Vinci did, the new biologism sees the universal body as a marker of inclusivity and readability, with deviations from that part of the unreadable landscape of uncertainty and risk. To develop the discussion further, attention now turns to exploring the ways in which biometric technologies map onto the geographies of disability.

Global Mobility and Difference

Massey's (1994) early ideas on mobility and unequal power-geometry are extremely useful in acting as a corrective to overly celebratory accounts of global mobility, multicultural enrichment and playful hybridity which were endemic during the 1990s in many academic accounts of globalisation and deterritorialisation. The post-modern turn in geography generated many overly celebratory liquid narratives of travel and flight that seemed to render the body free and weightless. It also spawned nomadic theories which too often ignored the gendering and racialisation of (im)mobility (see for instance Derrida's extensive work on the spaces and flows of postmodernity). In the face of the complex hypermobilities (the business traveller) and incarcerated immobilities (Guantanamo bay, ghettos and slums) we experience as part of globalisation, mobility scholars then, are arguing that we should pay much more attention to the often taken for granted meanings inscribed within all sorts of local and global human and material mobilities and immobilities within everyday life.

Immigration policy is a key way in which the state controls membership of its polity by deciding who can and cannot enter. And biometric technologies are increasingly being employed as a tool of surveillance which determines mobility and constructs barriers around and within nation states. This is particularly relevant in disability research *as constraints on movement have been fundamental to the oppression of disabled groups.* Indeed, 'barriers have connotations with physical space or obstructions such as walls, fences or other demarcations which prevent people having ease of access from one place to another' (Imrie 2001: 232). As such geographers must engage with the software sorted geographies and mobility of digital information that is increasingly creating society through the automatic production of space (Graham 2005). By space production being categorised as automatic, Graham is referring to the conscious moral withdrawal of human judgement, empathy and monitoring to a world in which the conditions, laws and ideals are automatically set and generated by technology. We see this often, through the automisation of many of the services we take for granted – automised check-outs, recorded telephone systems, automated train announcements and bank cheque paying in systems. However, the move to automise space production is not merely the drive to be efficient, but can also be a conscious desire of those in power to 'normalise' that power, by embedding decision-making processes within systems which have no affective registers as a means of abdicating moral responsibility, usually for economic gain.

Agamben's theories (1998) of exception and *Homo Sacer* have come as a welcome corrective to postmodern accounts of unbridled mobility and seem to have an extraordinarily wide purchase in explaining numerous areas of governance and everyday life, including the governance of mobility at many scales. To Agamben *Homo Sacer* represents bare life, a state of exception which exists in a liminal threshold between the spaces of law. His key insight is the ways in which democratic liberal governments are becoming totalitarian states through their powers of exceptionalism. Agamben highlights how these exceptional states are no longer temporary or fleeting but seem to have become the rule. Of course, there are many useful ideas within Agamben's work. For instance, there is much to be learned for mobility studies, in understanding how the concentration camp, the ultimate sphere of exception is also the ultimate space of violently forced immobility and thus ultimate power. However, Agamben does not really deal with issues of subjectivity, identity, race and disability in his work, and it is these two sets of ideas that we wish to connect here.

One of the main weaknesses is Agamben's assumption of *Homo Sacer* as an undifferentiated, interchangeable, disembodied, male figure that runs throughout the text, which limits his argument in profound ways (Mitchell 2006). To look more closely at the politics of exclusion, we need to examine how the early ideas around exclusion were crystallised in the late 1800s and throughout the early 1900s through supposedly scientific study of race and rationality in the formation of immigrant identities (Hannah, 2000).

Drawing on the work of Butler, Mitchell suggests that exceptions to the ideas of universalism do not occur only through the hierarchical production and maintenance of the 'norms of reason' as Agamben would have it; they are also simultaneously structured through the body especially through processes of sex, gender and race formation specific to the naturalised, biologised body of modernity. The problem is that Agamben does not sufficiently engage with how biological differences of race, gender and sexuality to name a few play into his theories of exceptionalism. This is something of an oversight given that, as Mitchell suggests, the ontologising of specific groups as scientifically distinct in their biology is precisely what makes them rationally accepted embodied 'exceptions' to the norm and thus exceptional.

The aim of this chapter then is to take the body, the geography closest in, as the key scale of analysis (an often forgotten scale of analysis), in order to try to flesh out how an awareness of the embodiment of exceptionalism can help us understand precisely how sovereign exceptionalism and legal abandonment now operates with such impunity and without moral contradiction in contemporary society. The ontological linking of race, gender, science and citizenship from at least the 1800s onwards was a key manoeuvre in allowing many kinds of states of exception to flourish whilst at the same time not seeming to undermine the legitimacy of the neoliberal project. Agamben does not fully explore how some human beings are ontologically produced through scientific and technological knowledge as different and inadmissible with abnormality literally inscribed in and on their bodies in such a way that they cannot be considered interchangeable with others. He fails to elucidate the way these corporeal markings create 'place-bound bodies' in contrast to the global neoliberal norm of the placeless and free floating body. In the remainder of this chapter then, a case study is presented to explore the new geographies that are being automatically created through the introduction and diffusion of biometric technologies such as facial recognition, fingerprinting and iris scanning at border checkpoints.

Biometric Technologies and Citizenship

Within disability studies, research has illustrated the ways in which certain bodies are situated outside the norms that form the basis for citizenship (Davis 1995). Whilst biometric systems would appear to equally affect all citizens, regardless of nationality, the majority of biometric developments have disproportionately focused upon non-nationals. Biometrics can be regarded as an important tool used to include or exclude individuals from the nation's polity and citizenship. Indeed, citizenship, medical screening and surveillance technologies already operate in countries such as Canada to separate qualified, worthy citizens from unqualified, unworthy lives (Weibe 2008). State strategies are thus employed to screen potential citizens based upon physical and mental health and result in a form of demarcation that operates as a type of discrimination and form of inequality. Surveillance operates as a technique of social control to document and classify people in an attempt to force a population to conform to social norms. For example in Canada some residents with certain illnesses must undergo surveillance (Weibe 2008). Similarly biometric technologies serve to operate as a technology of surveillance implemented by the state to determine access to the nation state and citizenship rights. Just as the unhealthy body has been deemed unfit by certain governments and resulted in the mobility of individuals or families being constrained, inherent biases and potential abuses of biometric systems may impact disproportionately upon bodies that do not conform to the norm, with resulting impacts upon access to rights. This could lead to a situation where only those individuals with the 'right' biometrics will have access to public services. Indeed, it has been theorised that bodies 'will become inscribed with their identities as citizens, with all the rights and benefits that status brings along…the citizen's body thus will become implicated in the distribution of benefits, services and rights' (Van der Ploeg 1999: 296).

The importance of biometrics relative to entitlement to state goods has surfaced in relation to ID cards. It has been suggested that legislation being implemented to introduce ID cards may result in vulnerable groups, like asylum seekers, being denied essential services such as health care. Whilst plans for ID cards are not fully operationalised the use of biometric identifiers to determine access to socio-economic goods is already determining the treatment of asylum seekers in the UK. Asylum seekers have been historically issued with the Standard Acknowledgement Letter (SAL) but this was considered inconvenient or not robust enough and open to counterfeiting and forging. As a result asylum seekers are now issued with an Application Registration Card (ARC). This is the Home Office smart card for asylum seekers which is intended to provide fast and positive identification of applicants subsequent to their initial processing at ports of entry or at the Croydon Asylum Seekers Unit. But above all, this system has been designed to stop asylum seekers who may be doing unlawful things such as applying twice for benefits (Gardner 2002). The production of biometric identities (the creation of state knowledge) is therefore bound up with citizenship rights and access to state goods. Indeed, 'biometrics seem to be about maintaining social order by regulating in- and exclusion from socio-economic goods, geographic spaces and liberties' (Van der Ploeg 1999: 296). And this is of particular relevance to disability scholars given recent debates on the right to receive state welfare in situations where individuals are physically impaired, such as Incapacity Benefit in the UK.

Biometric Design, Disability and the Body

Contemporary geopolitical concerns, the rapid growth in information technologies and the flourishing surveillance culture have all led to the increasing use of biometric technologies (Czegledy and Czegledy 2002). The combination of these factors has resulted in the 'technological separation of corporeal identity from personal bases of knowledge' (Czegledy and Czegledy 2002: 76). Biometric technology is thus a powerful tool of biopolitics; the supervision of the body that operates through a series of regulatory controls and surveillance (Legg 2005). In current immigration regimes one major feature is the mistrust of individuals entering national borders and the difficulties faced in determining the national or ethnic identity of migrants. It is no longer regarded as possible for immigration officials alone to ascertain identity by 'reading' this from the individual. One specific response to this has been the move towards placing trust in technological advances. This has led to a distinction being made between the 'individual' and the 'body'. The body is no longer just a biological object but redefined as 'text' which can be confidently and accurately read by advanced technology. The body is a source of 'truth' or 'authenticity' (Ball 2005). As a result, the decision reached by employing biometric technology is regarded as 'truth', which contrasts the potentially misleading information provided by individual migrants. Nevertheless, one has to question in designing biometric technologies what 'truth' is being sought. Imrie's (2003) seminal work on architectural design and theory highlighted the failure of architects to recognise bodily and physiological diversity with standards and dimensions revolving around the conception of a 'normal' body. In the same manner there may be concerns over the inherent design assumptions of biometric technologies, which will undoubtedly lead to potentially differential impacts upon sectors of the population, like disabled groups. So there is a need to unearth and challenge potentially ableist values that are etched within the fabric of biometric technology research and development.

Spatialities of disability are inscribed by biomedical discourses of the body, with discourses propagating a conception of disability as abnormal and deviant (Butler and Parr 1999). Appearance is a key modality by which disability is constructed, with the body of the disabled person marked by the disability (Davis 1995). Likewise, by means of biometric technologies the body is marked out as different. Interaction with biometric technology not only incorporates the body as a material entity but also inscribes it through marking and recording (Ball 2005). In creating a template or readable 'text' of the body, bodies are marked with a sign that can be read by the appropriate equipment wherever they go. Recent discourse surrounding the response to immigration has become increasingly linked to anti-terrorist legislation (Thomas 2005). So in addition to immigration policy being linked to the control or management of migration (or bodies) as a means to preserving national borders, rhetoric and rationale now equally focuses upon the terrorist threat posed by 'foreign' bodies. Theorists have argued that the 'black' body identifies individuals as being 'foreign' or 'other' (Sibley 1995). Bodily appearance is regarded as a visible marker of difference. In a similar manner, through the usage of biometric technology, the body comes to be marked with stigmata, signed on the flesh (Van der Ploeg 1999). It is notable that in criminal law, the taking of fingerprints is allowed only for suspects of serious crimes, who may be taken into custody or detention. Nonetheless, this activity is routine for asylum seekers entering the UK and legitimised by the rhetoric of maintaining security and ascertaining the 'true' identity of asylum seekers. The stigma of criminal activity inherently attached to fingerprints is attached to the refugee 'body'. Accordingly, the biometric fingerprint becomes the mark of illegality written on the body. As one recent commentator stated 'biometrics has always had a role to play in identifying the so-called "bad guys"…watch any number of modern-day thrillers or sci-fi movies and it won't be long before fingers and eyes are being chopped and popped for someone to bypass a biometric security system' (Lynch 2005: 1).

In sum, recent debates surrounding biometric geographies intersect with disability scholarship in multiple and complex ways. Questions remain as to how restrictions imposed by biometric systems interface with debates surrounding global mobility, citizenship and disability. Literature within disability geography has developed beyond medical to social models of disability to suggest ways in which 'debates and actions surrounding "disability" should shift away from concentrating on physiology in order to view more critically the society in which individuals and their impairments are contextualised' (Butler and Parr 1999: 4). The discourse of disability is said to include social, economic, historical and cultural processes that regulate and control the way we think about the body (Davis, 1995). Exploration of how biometric technologies are developed and their implementation is therefore vital in understanding the ways in which disabled groups may be potentially excluded and constructed as deviating from the norm. This chapter thus aims to expose ableist ideas, practices and institutions within the biometrics industry, which includes assumptions about ablebodiedness and constructing disabled persons as marginalised, invisible 'others' (Chouinard 1997).

Methodology

The first phase of the project involved desk based research to facilitate collation and synthesis of academic and policy papers. The second phase of the research focused upon primary data collection. During summer 2005 attendance at the International Biometrics Conference in London enabled first hand observation of the latest developments in biometrics technologies displayed in the exhibition centre. This also facilitated meetings and interviews with high profile contacts in the biometrics industry, both users and producers of a range of biometric systems. The producers included companies that develop, manufacture, pilot and 'test' biometric technology. The users in this context are the authorities such as ports/ airports which implement biometric technology at borders. During the project a total of ten key informants were interviewed both in London and elsewhere in the UK. This consultation period with key actors served to clearly outline the framework of the biometrics industry whilst also providing in-depth understanding of how biometric technologies are developed and implemented.

Discussion

In relation to biometrics there is a distinction made between deterministic and voluntarist views of technology (Van der Ploeg 2003). The deterministic view essentialises technology, reifying it and considers it as a stable object. However, in contrast the voluntarist view stresses technology as a human practice, informed by economic, cultural, normative and social factors. This means technology is regarded as non-stable and fluid. First, amongst our respondents there was evidence of technology being essentialised and defined in scientific terms:

> If you think of the stripes in the human iris…these stripes are there and they can be thought of as a circular bar code that holds enough codes enough bar codes, thick and thin stripes, it could hold enough codes to encode everybody in the universe who's ever been born. Not just planet earth! Incredibly large numbers of people can be encoded and never be confused. But the failure to enrol is partly a human factor, people dislike prolonged enrolment (Biometric technology producer).

> The matching process is a specific mathematical formula that is enabled by the biometric software (Biometric technology producer).

As noted by the second respondent, biometric technology is defined in mathematical terms conveying a sense of objectivity and infallibility. Similarly the first respondent defines the operation of biometrics in terms of coding the human iris. Alongside this discussion of scientific rigour, however, there is reference to problems that may arise. Interestingly, however, the interviewee attributes such issues to the 'human factor' and fails to question the infallibility of biometric systems themselves. The body is conceptualised as failing to meet normal standards and considered to be at fault (Butler and Parr 1999). Thus, despite attempts by producers to emphasise the effectiveness and accuracy of the emerging biometrics market, recent critiques of facial recognition systems have suggested that a 'biology of culpability' often acts to unwittingly embed prejudice into code where it is then shrouded in an aura of scientific objectivity (Rose 2003). Numerous trials have identified a range of technical problems associated with the technology and its use on particular social groups and populations. Amongst the respondents the infallibility and objectivity of biometrics was questioned specifically in relation to its implementation and operationalisation:

> In photographing the iris, in some cases it just wouldn't accept (Biometric technology user).

> The face machines I looked at, all shows quite a lot of promise when demonstrated but when I tried them I found that if I had my haircut or put on spectacles the thing just fell to the ground (Biometric technology producer).

> I suppose what I'd say is that one size doesn't fit all (Biometric technology consultant).

Beyond the general failure of systems to enrol individuals or recognise change, there are specific groups that are more likely to challenges biometric systems. A UK passport service enrolment trial highlighted difficulties in enrolment for those with physical disabilities who may have difficulties in positioning themselves correctly for enrolment. Other studies have highlighted lower enrolment and verification success rates for a variety of social populations (most notably those with dark brown eyes, black people, the elderly, young females, tall people, twins and those with highly similar facial features). This mirrors other reported trials of facial recognition by CCTV, which claim success rates that vary considerably depending on the gender, ethnicity and age of the subject as well as a whole host of other environmental, geographical, cultural, contextual and postural factors (Graham 2005). These issues were recognised by our respondents:

> For someone in a wheelchair if you can't perfectly adapt your position it could be difficult. For blind people it certainly can be difficult because they can't see… You don't actually have to focus, but you do have to keep a constant relationship with the camera…that's why we couldn't get acceptable enrolment (in a recent trial) for quite a large section of people with disabilities (Biometric technology user).

> With iris recognition for example, we had about a 90 per cent success rate for our quota sample – for our representative sample and only about a 60 per cent success rate for our disabled group…I would say the difficulty we had with iris was positioning…Iris recognition does require you to keep a focus for a certain length of time and the position between you and the camera is quite critical. Now for most people it's fine…though for someone in a wheelchair if you can't perfectly adapt your position it could be difficult (Biometric technology user).

It would seem, then, that popular technopolitical understandings of the body and the biographical identity implicitly accept the nineteenth century notion of the body and identity as solid, stable legible entity, rather than as something mutable, fluid, fractured and contingent (Cole 2001). The problem in the design of biometric technologies is that they 'reduce bodies to a singular type, in which spaces and the bodies that populate them, are neither sexed nor gendered, nor racialised or inscribed by class' (Imrie 2003: 50). Similar to the architects in Imrie's (2003) research, the design of biometric technologies appears to be premised on the notion of a widely applicable image of a white, able-bodied, masculine body. Bodily conceptions of fit and healthy bodies dominate the assumptions of biometric technologies. As such, if bodies fail to conform to this image, then this disrupts the system. Indeed, fluidity and temporal changes create particular challenges for biometric systems as evidenced below:

> Fingerprint would be one example, where the fingerprints fade, old people tend to get dehydrated, so fingerprints are easier and there are also certain races and nations where people are very difficult to recognise by any given single biometric (Biometric technology producer).

There are radial stripes (iris) born in the foetus and stable once a person is adult, bit unstable in teenagers apparently, that's another problem – teenagers! (Biometric technology producer).

And people who are regularly doing manual work, like brick layers in London say, the cement, alkaline in cement dissolves the surface of the skin and they can get cracks on their fingers and there are other reported instances of people pretending to be disabled by sandpapering their fingerprints (Biometric technology producer).

It would seem then, that there is a resurgence of biologism despite the promise of equality offered by post-modern theory. Of course, in reality, we know facial measurements are likely to change throughout the life course, through a change in weight, through the ageing process or even through cosmetic surgery. It is no coincidence that those features that Derrida calls points of recognition, the eyes and hands, which are able to express the dynamism of the internal self, are the hardest to 'fix' into an immutable mobile.[1] Nevertheless, it can be the minutiae of human activity that can create problems for the technology. For example, a smile causes havoc for a biometric identifier with no sense of humour:

To minimise error rates, there is a need for a rigorous standards of electronic images. Smiles disrupt the use of facial recognition technology. They cause an earthquake in terms of facial recognition (Biometric technology consultant).

So whilst it may be easy to subvert the technology either on purpose or unwittingly, it's far less easy to resist the underpinning ideologies that govern the production of the technology that see the body as fixed, stable, solid and individual and identity as being biographically continuous. For a start it is a technology that will not recognise difference, diversity and anomaly – the scratched fingerprint of the rock-climber, the ageing iris of the pensioner, the fingerprint-less fingers of a young boy with a genetic disorder or the veiled face of a Muslim woman in a Hijab. In a worse case scenario, the state through a process of subjectification literally sets the boundaries of action, of being in the world through its vast database of globally interoperable information that it uses to track individuals as they move around the world in a seen, legible uniformity of action and purpose. Furthermore, biometric screening systems increasingly create a time differential between the so-called 'kinetic elite' and less favoured passengers, creating a global othering process so that eventually the migrant has nowhere left to go/hide and becomes increasingly marginalised spatially and symbolically. As one respondent stated, when asked about the purpose of biometric technology:

---

1   Derrida discusses the idea of the eyes as dynamic in the film *Derrida* directed by Kirby Dick and Amy Ziering Kofman and released in 2002 (Zeitgeist films).

To get a clear idea of who people are, and to eliminate people on lists that you
do not want…negative identification means eliminating an individual from a list
of 'undesirables' (Biometric technology consultant).

This double tracking includes dual embodied exceptions – the global elite, political
leaders, military personnel and business people for instance, are increasingly to be
biometrically coded as free from the particularities of place and body, exempt
from the normal laws of society in business, at the border and at war. This is in
stark contrast to those screened at borders and declared unfit for rapid passage, or
immobilised completely at detention centres and immigration prisons. Thus 'in
the name of security, symbolic violence is practised in the form of internalized
humiliations and legitimization of inequality and hierarchy connected to biometric
identification' (Häkli 2007: 140).

Where do individuals with disabilities sit in this continuum? The empirical
results are still to be built upon, however, it seems that as a generalisation, disabled
groups may be harder to biometrically identify, because of the way the software has
been programmed to look for consistencies. This is important because as authors
such as Graham have noted, those most vulnerable in society may be precisely
those that do not have an accurate biometric identifier. A stereotype image of this
is Tom Hanks' character in *The Terminal*, who cannot step outside the airport
without being arrested. However, in the spaces between nation-states, this may not
be too far from the truth. The biometrically inadmissible are caught in the liminal
non-spaces of mobility such as the airport, the interrogation room and detention
centre. Ironically, it is these very spaces of exception that allow the contemporary
global hierarchies of inequality to persist in the face of the contemporary spectre
of migration of the masses alluded to by Hardt and Negri (2000) in *Empire*. Our
argument in this chapter is that technological producers need to think more carefully
about disabled users in constructing and rolling out these technologies, to avoid
those with disabilities becoming conflated with other 'inadmissible bodies', and
to think more deeply in general about the way in which inadmissibility is defined
in the contemporary era.

Conclusion

To conclude, the foundations for current technologies were set down in the
progressive era when bodies and identities were thought of in a fixed, deterministic
way (see for instance, Maddern and Desforges' (2004) work on the eugenics
movement at Ellis Island Immigration Station). As we moved towards the feminist
and civil rights movements of the 1960s and the anti-racist movements of the
1970s, the promise of postmodernism was one of a world in which difference
and deviance from so-called 'normality' was not only acknowledged but also
celebrated. However, this chapter argues that with the rise of fears founded in
an ideologically-based warfare and terrorism, biometrically scientised codes

of embodied 'difference' have created a new type of eugenic thinking which unwittingly reproduces the ontological divisions between ethnicities and disabled populations founded in the modern era of territorial expansion and protectionism, thereby becoming the new technologies of biopolitical control governing both individuals and populations. The chapter has exposed the ableist assumptions that permeate biometric technologies, which has important implications for disabled persons that are constructed as being marginalised, invisible 'others'. There is a need to challenge this (re)new(ed) hegemony of 'normalcy' which dominates society and which evidently permeates the biometric arena (Davis 1995).

Developers and consultants in the biometrics industry, whilst technologically advanced, do not always embed the technologies in principles of social egalitarianism. Where the industry lags behind, literature on the geographies of disability highlights several fruitful ways to progress. Perhaps one way forward is to draw upon 'accommodationist' agendas which highlight the need to be alert to differences in the 'doing of things'. Hansen and Philo's (2007) work on the normality of doing things differently highlights the insistence to register the multitude of ways in which different bodies do (and can do) things differently. Finally, by viewing Agamben's ideas of exceptionalism through the lens of the biologised body of modernity we can seek to fully understand, document and theorise these double tracked global dualities of inclusion and exclusion. Whilst academics and activists cannot easily resist the control creep engendered by biometric technologies, we would like to argue that beyond the biopolitics of neo-liberalism, there is a 'biopoetics' of hope, which comprises the experiences of those who grow, move, change, travel, and make human connections. By this the authors suggest that the biopoetics/biopolitics dualism mirrors the theoretical distinction that has been made between disability being seen as a 'medical' issue versus disability being seen as a 'social' issue. Thus rather than seeing the body as a text that can be read by machines, a body as a map of a set of 'symptoms' of guilt or innocence, of conformity or deviation, or Vitruvian physicality (or a lack of), biopoetics looks at the body in a different way – as a unique embodied space of narratives and potential. These are the poetics of embodiment that are not reducible to uniform lines of numerical code. By collecting these biopoetic narratives in stories such as the one we have just begun the task of providing and interpreting here, academics and activists can provide a counterbalance to the biopolitics of neoliberalism.

References

Adey, P. 2004a. Surveillance at the airport: surveilling mobility/mobilising surveillance. *Environment and Planning A*, 36, 1365-1380.

Adey, P. 2004b. Secured and sorted mobilities. *Surveillance and Society*, 1(4), 500-519.

Agamben, G. 1998. *Homo Sacer: Sovereign Power and Bare Life*. Stanford: Stanford University Press.

Agre, P. 2001. Your face is not a barcode: arguments against automatic face recognition in public places. *Whole Earth*, 106, 74-77.

Amoore, L. 2006. Biometric borders: Governing mobilities in the war on terror. *Political Geography*, 25, 336-351.

Ball, K. 2005. Organization, surveillance and the body: towards a politics of resistance. *Organization*, 12(1), 89-108.

Bashford, A. and Strange, C. 2002. Asylum-seekers and national histories of detention. *Australian Journal of Politics and History*, 48(4), 509-527.

BBC News 2008. Arrests made at ID cards meeting. *BBC News Online*, 4 July 2008.

Beck, A. and Broadhurst, K. 1998. Policing the community: the impact of national identity cards in the European Union. *Journal of Ethnic and Migration Studies*, 24(3), 413-431.

Bevan, G. 2008. ID Scheme: the truths, half-truths and deceptions. *The Herald*, 4 July 2008.

Butler, R. and Parr, H. (eds) 1999. *Mind and Body Spaces: Geographies of Illness, Impairment and Disability*. London: Routledge.

Chouinard, V. 1997. Making space for disabling difference: challenging ableist geographies. *Environment and Planning D: Society and Space*, 15, 379-387.

Crooks, V.A., Dorn, M.L. and Wilton, R.D. 2008. Emerging scholarship in the geographies of disability. *Health and Place*, 14: 883-888.

Czegledy, N. and Czegledy, A.P. 2002. The body as password: biometrics and corporeal dispossession. *Filozofki vestnik*, XXIII(2), 75-92.

Davis, L.J. 1995. *Enforcing Normalcy*. London: Verso.

Gardner, G. 2002. Biometric smart identity card for UK asylum seekers. *Biometric Technology Today*, January, 2.

Graham, S. 2005. Software-sorted geographies. *Progress in Human Geography* 29(5), 562-580.

Häkli, J. 2007. Biometric identities. *Progress in Human Geography*, 31(2), 139-141.

Hannah, M. 2000. *Governmentality and the Mastery of Territory in Nineteenth-Century America*. Cambridge: Cambridge University Press.

Hansen, N. and Philo, C. 2007. The normality of doing things differently: bodies, spaces and disability geography. *Tijdschrift voor Economische en Sociale Geografie*, 98(4), 493-506.

Hardt, M. and Negri, A. 2000. *Empire*. Cambridge, Mass: Harvard University Press.

Imrie, R. 2001. Barriered and bounded places and the spatialities of disability. *Urban Studies*, 38(2), 231-237.

Imrie, R. 2003. Architects' conceptions of the human body. *Environment and Planning D: Society and Space*, 21, 47-65.

Joint Committee on Human Rights 2005. *Identity Cards Bill*, Fifth Report of Session 2004-05, House of Lords, House of Commons, HL Paper 35.

Legg, S. 2005. Foucault's population geographies: classifications, biopolitics and governmental spaces. *Population, Space and Place*, 11(3), 137-156.

Lynch, M. 2005. Identity crisis. *VNU Network*, available at: http://www.vnunet. com/print/it/1160571.

Maddern, J. and Desforges, L. 2004. Front doors to freedom, portal to the past: history at the Ellis Island Immigration Museum, New York'. *Social and Cultural Geography*, 5(3), 437-457.

Malmberg, M. 2004. Control and deterrence: discourses of detention of asylum seekers. *Sussex Centre for Migration Working Paper*, 20, University of Sussex, Sussex Centre for Migration.

Massey, D. 1994. *Space, Place and Gender*. Minneapolis: University of Minnesota Press.

Mitchell, D. 2002. Controlling space, controlling scale: migratory labour, free speech, and development in the American West. *Journal of Historical Geography*, 28(1), 63-84.

Mitchell, K. 2006. Geographies of identity: the new exceptionalism. *Progress in Human Geography*, 30(1), 95-106.

Petermann, T., Sauter, A. and Scherz, C. 2006. Biometrics at the borders – the challenges of a political technology. *International Review of Law Computers and Technology*, 20(1-2), 149-166.

Philo, C. 2001. Accumulating populations: bodies, institutions and space. *International Journal of Population Geography*, 7, 473-490.

Sibley, D. 1995. *Geographies of Exclusion*. London: Routledge.

Thomas, R. 2005. Biometrics, international migrants and human rights. *Global Migration Perspectives*, 17 January 2005, available at: www.gcim.org.

Van der Ploeg, I. 1999. The illegal body: 'Eurodac' and the politics of biometric identification. *Ethics and Information Technology*, 1, 295-302.

Van der Ploeg, I. 2003. Biometrics and privacy: a note on the politics of theorizing technology. *Information, Communication and Society*, 6(1), 85-104.

Wiebe, S. 2008. Opinion. Rethinking citizenship: (un)healthy bodies and the Canadian border. *Surveillance and Society*, 5(3), 334-339.

Chapter 14
Geographies of Disability:
Reflections on New Body Knowledges

Isabel Dyck

Javier is only a week old, but already he is at the centre of a growing political storm. In the first case of its kind in Spain, he was born through stem-cell selection because his parents hope that he will provide a cure for his chronically ill brother, Andres. The six-year-old suffers from a congenital form of anaemia, but material from Javier's umbilical cord may help to cure him because screening has ensured that his new brother is free of the condition. However, the method of Javier's birth has brought criticism from the Roman Catholic Church and 'pro-life' groups, who condemn stem-cell screening.... [the] director of the Life Foundation, said that the start of a new life was always good news, but added: 'The method of this birth is degrading for human beings to have been selected like a prize' (*The Times* 2008a: 47).

This news item captures the complex threads weaving through contemporary bioethics: advances in medical knowledge and biotechnology; social and faith attitudes and values; the personal pain and hopes of a particular family; and – less prominent in this article – distributive inequalities across the globe. My aim here is not to step into the labyrinth of bioethics, but rather draw on these threads in reflecting upon the contemporary context of research in the geographies of disability – a field where politics, social values and personal experience have never been far from view. One dimension of this politics has been the incorporation of the body in theorising disability. But what body for what politics? Certainly the biosciences and medicine have transformed our understanding of the body, with, I suggest, implications for how we incorporate this 'medical' body into our research approaches. It is now some time since a rapprochement with the medical has been encouraged, in recognition that the discourses and practices of medical knowledge are necessarily intertwined with the conceptualisation, management and experience of 'disability' (see Parr 2002 for a review). Recent biomedical and technological advances challenge the agenda of disability geographies further. Here I consider some of the implications – selective in scope – of such advances in 'body knowledge' for research in our sub-discipline.

Disability is a complex site of theorising, with its politics, bodies and power relations well rehearsed. Ways of conceptualising disability have revealed an array of thinking about how best to address its contextuality and the varied experiences of disability or chronic illness. The meanings of disability, as long recognised, shift and have different consequences over time and cross-culturally, with political economy and welfare regimes having profound effects for those deviating from the able-bodied norm (Gleeson 1999). Some of the most powerful writing on such issues has incorporated an autobiographical dimension, such as the seminal work in sociology of Michael Oliver (1990) on the social model of disability and ensuing feminist critique, such as that of Jenny Morris (1996). In geography there has been a similar strand of writing from experience – from a variety of theoretical perspectives – building on work signalling the ableist society and disabling environments in which we live (e.g., Chouinard 1997).[1] I have not written from such experience although my work is indebted to those for whom disability and/or chronic illness is central to the navigation of their lives.[2] Further, it is in reference to the rehabilitation science students I taught for some years that the 'medical' body has vied prominently with theorisations of the body that increasingly point up the inseparability of the material and representational body in understanding the embodied nature of disability experience. Such embodiment has a geography, intertwining with particular temporal and spatial contexts intimately connected to the spaces of everyday life – the home, neighbourhood, workplace and the myriad of institutions and state agencies that directly or indirectly intervene in lives.

Located within the tension of teaching future rehabilitation professionals working in the shadow of biomedicine and a research programme based on critique of a biomedical framing of health, illness and disability issues, I have found an interesting, epistemological space through which to consider issues of embodiment. Its contradictions, tempered with common working/research goals of addressing access issues – albeit through different practices and frameworks of understanding – have provided a rich ground for considering the relationship between theoretical models and the attendant politics and ethics of research in constructing geographies of disability. I have been particularly interested in the interconnections between theorisations of the body, identity, gender and environment in understanding how disabled people and those living with chronic illnesses habituate and move across the spaces of everyday life and how these spaces, lived experience and subjectivity are recursively shaped. This interest is echoed in a variety of ways in disability geography, a range of work that shows that conceptualising 'context' or 'environment' is far from simple (e.g., Butler and Parr 1999, *Environment and Planning D* 1997, Gleeson 1999, Kitchin 2000). Yet conceptualising the environment has been crucial to the politics of disability research in delineating issues of access, a crucial dimension of a socio-spatial model of disability.

_____

1   See Imrie and Edwards (2007) for a review of contemporary work and issues in geographies of disability.

2   In particular I am indebted to the rich experience of working with Pamela Moss on theorising the notion of embodiment in the context of women with chronic illness and conversations with anthropologist Beverley Gartrell, in the context of her personal and professional life.

Until relatively recently, the body has received less attention than the environment. But an increasing call for integrating the body in disability research (Hall 2000), and research that 'takes the body seriously' shows that how the body is integrated into models of understanding is critical to the political freighting of research. Take for example three approaches. First, the reductionist conceptualisation of both bodies and environment in rehabilitation models, which while aiming to be holistic and sensitive to the individual and his or her environment retain an untheorised body and a static conceptualisation of environment. Definition of the environment may include discussion of disability policy relations and some practitioners may engage politically as disability activists in their personal lives, but analysis tends to be devoid of theorised power relations, with normative concepts upheld. Second, the social model of disability's focus on physical and attitudinal barriers – the disabling features of an ableist society – was to the neglect of the fleshy, corporeality of the body. The neglect was strategic, in fear of the spectre of biological essentialism muddying the waters of the social model's political message, which put the socially constructed, disabling environment at the centre of pursuing social justice. While more nuanced accounts of disabling spaces have followed in response to critique of an earlier rendition of the social model – many from geography (see Imrie and Edwards 2007) – proponents of the social model of disability and activist groups are still wary of incorporating the biology of the body (notionally framed as impairment) in theorising disability. Third, and in contrast, post-structuralist informed accounts have embraced the body, centring the concept of embodiment in drawing connections between relations of power and lived experience. In such an approach the experience of corporeality cannot be isolated from the intertwining cultural codings of bodies and city spaces (cf. Imrie 2001), as well as the 'hard facts' of medical science.

The political work these different 'bodies' do is well rehearsed. The first works with the taken-for-granted, hegemonic medical body – its anatomy, physiology, biology at the point of medical and social welfare intervention while it recognises that people live within the particular homes, neighbourhoods and networks of social relationships; the second, underpinning the social model, conveys the deliberate politics of a disability activist community/collective voice; while the third (an intertwining material/discursive body) informs the 'body politics' first signalled by Dorn and Laws (1994) and latterly developed as the 'radical body politics' of a feminist materialist perspective (e.g., Moss and Dyck 2002). The tension of the politics between different approaches to theorising disability and its bodies in geographical work is interestingly signalled in the sometimes uneasy alliance of disability and chronic illness research. Yet I see similar processes at play in the social and economic marginalisation of both people with stable conditions and those living with the fluctuations of chronic illness. Research with a focus on the embodied experience of disability/chronic illness has challenged binaries, such as health and illness, able and disabled, showing the dynamic nature of such categories. This work suggest a politics of the body that recognises the temporality and situatedness of chronic illness and disability experience, and also alerts us to the political and social construction categories that shape access to resources (e.g., Crooks and Chouinard 2006, Hansen and Philo 1997, Moss and Dyck 2002).

There is now a rich vein of work that pays attention to embodiment in various ways in disability research, but as social scientists have been refining theories of the body, so too has science been developing knowledge about bodies, notably in the field of genetics, which poses challenges for how disability researchers incorporate the body in analyses. Social and technological changes of the twenty-first century further complicate how the effects of new body knowledges form a context of disability and chronic illness experience and body politics. These new ways of seeing and understanding of the body go beyond the 'clinical gaze' identified by Foucault (1973) to another scale – that of the molecular level. Moving to this level of body knowledge spills over into public health discourse, disability discourse and ethics, further blurring a distinction between disability and chronic illness as categories of research – rather these can be viewed as parallel theoretical sites. The new *certainty* of the body wrought at the molecular level also raises new *uncertainties* in terms of associated action, with potentially profound implications for disability research and its politics. As Rose (2007) and Agamben (1998) alert us, new knowledge and new technologies are taking us into a realm where the value of 'life itself' is re-questioned – whose life, what life is to be valued? And who is to manage 'life itself'?

Drawing on Fludwig Fleck's notion of 'styles of thought', Rose (2007) argues that the advancement of medical knowledge to the level of the molecular is not, however, just a matter of new knowledge; rather it leads to seeing objects in new ways – providing a new way of understanding WHAT there is to be seen, including life:

> techniques of gene cutting and splicing, the polymerase chain reaction for creating multiple copies of precise segments of DNA outside living systems, the customized fabrication of DNA sequences to order, the manufacture of organisms with or without specific gene sequences...new forms of molecular life. And in doing so, it is fabricating a new way of understanding life itself. (Rose 2007: 13)

This realm of knowledge and understanding is intimately connected to the social world – it is not knowledge in a vacuum, but rather knowledge intimately involved in the constitution of human subjectivity and the regulation of populations. As Rose (2007: 4) comments, the 'vital politics' of this century 'is concerned with our growing capacities to control, manage, engineer, reshape, and modulate the very vital capacities of human beings as living creatures', a capacity that is associated with a proliferation of 'body experts' (subprofessionals with somatic expertise) involved in the governing of human conduct. In parallel, human beings employ a new sense of ownership of the body, which includes not just rights to a cure but an orientation towards a goal of achieving the best possible life. Further, current biomedical knowledge and techniques that allow us know life at the molecular level spill over into a new focus on optimisation of the body in public health, now an arena where self-disciplining forms of biopower can be side-stepped as 'body projects' can involve technoscience that can directly alter and enhance 'the material stuff of life itself' (Stepnisky 2007: 195).

Yet the ultimate responsibility for managing the vital politics of the body is left to the individual. This is not new, for critical public health work has noted that the management of 'bodies at risk' involves a moral imperative and responsibility on the part of individuals to keep healthy (Brown and Duncan 2000, Petersen 2007). But the extension of that social responsibility from simply achieving health to maximising the capacity of life brings a new sense of 'biological citizenship'. This latter incorporates ideas of human subjectivity and 'life' that lead to a recoding of rights and responsibilities in relation to how life should be conducted – with a focus on corporeal, bodily existence (Rose 2007). So going back to Javier and his family's situation, with which this chapter opened, we can see this biological citizenship and the imperative of producing the best possible life – with the use of available technologies – in play. Yet, the matter cannot rest there. There is clearly much controversy in the use of new technologies around life, particularly in terms of groups that have other values concerning 'life itself'. Faith groups, as the news article indicates, provide one voice that questions the manipulation of life. But what of disability advocate groups and researchers? There is much at stake for those who have fought hard for the acceptance of 'difference' and the erasure of disabling environments.

The replacement of health/illness as the cornerstone of debate and action by a notion of optimising life has considerable import for thinking through disability and chronic illness and the regulation of non-normative bodies. But as Shakespeare (2005) comments there is far from a united voice within disability research and activism. One dimension of divided opinion is the difficulty of distinguishing between therapeutic intervention and enhancement of the normal (cf. Buchanon et al. 2000). At what point is a biological predisposition or characteristic to be corrected or eliminated? Is facial surgery for children with Down's Syndrome to be welcomed or abhorred? Petersen (2007) raises the futuristic scenario of 'designer babies', with genetic selection based on desirable characteristics giving people an imagined advantage in life. New possibilities for engineering appearance are paralleled by shifting social norms of the 'appropriate' body, while western discourses of the 'ideal' body are spreading globally. Already surgical intervention in Asian populations – for example, leg lengthening to enhance job success and rounding eyes – suggest 'better than average' bodies are modelled on notions embedded in political economies and cultural ideals of the global north (e.g., Kaw 1993). Yet at the same time, changing attitudes to disability and the chance for a good quality of life due to improved resources are posited as reasons for more babies with Down's Syndrome being born in the UK (*The Times* 2008b).

Shifts in social attitudes and improved accessibility have been hard won, and Shakespeare (2005) notes that there has been little engagement by disability writers with discussions of preventing impairment in pushing social justice issues.

As he states, the reluctance to engage with the issue of preventing impairment can be contextualised in the devaluation of disabled lives and the focus on biology that underlies genetic approaches. Genetic testing for potential impairment or chronic illness has particular purchase in debate, as genetic knowledge in this context raises the spectre of biological essentialism, as did the fleshy, corporeal body in the social model of disability. At the same time, screening provides a parent the opportunity (and responsibility) of deciding whether to continue a pregnancy where poor quality of life might be anticipated (Browner and Press 1995). While Shakespeare acknowledges the anxiety over the possibilities of 'eugenics by the back door' and detraction from discrimination as a major force in the social construction of disability, he believes a neglect of biomedical approaches is mistaken. He argues that biomedical cures and prevention of impairment are important for disabled people, citing evidence that many disabled people with degenerative conditions would welcome prevention or cures, and would be supportive of gene therapy or stem cell research. He supports the treating and reducing instances of congenital impairment when interventions are appropriately administered and where there is an absence of prejudice against disabled people. For Shakespeare the primary issue is that of distributory justice, related to inequalities of access to services both within a state and globally. This can be argued on different lines, but the fundamental issue is:

> ...that problems with genetic research and intervention often arise from political contexts and consequences, rather than intrinsic moral questions or the messages about disability which such genetics may be sending (Shakespeare 2005: 94).

In addition to access issues, there are considerations related to the global circulation of expertise, technology, patients and body parts implicated in potential intervention in disability and chronic illness states. Inequality issues are at the heart of such 'solutions'. The search for surgical enhancement is creating a medical tourism from affluent to less wealthy countries where the majority of the local population does not have access to procedures on offer (see Chinai and Goswani 2007). A market in body organs to enhance or prolong life of some includes the sale of organs by the very poor to sustain basic living needs (Cohen 2003, Nullis-Kapp 2004, Scheper-Hughes 2000). Certainly there are 'new' geographies to discover in an era of body optimisation. Rose (2007: 17) stresses that technologies of optimisation do not, however, consist simply of equipment or techniques, but rather are 'hybrid assemblages of knowledges, instruments, persons, systems of judgement, buildings and spaces, underpinned at the programmatic level by certain presuppositions and assumptions about human beings'. Our discipline is well placed to pursue the geographies of such technologies, whether in terms of distributive justice or the spatiality of processes involved in their adoption and effects.

We need to pay attention to the sites and spaces of the new techniques of governmentality as 'optimisation' is put into practice through a range of technologies. The (ab)normality of bodies is underpinned by social values, reflected in classificatory systems. In what ways are 'technologies of optimisation' involved in processes of normalisation and at what scales? For instance, how are new ways of knowing the body and technological advancement incorporated into normalisation processes, altering subjectivity and our relationship to 'expert' knowledge? These are critical questions to address if we are to understand how changes in discursive representations of normativity become embodied and constituted through socio-spatial processes. If conceptions of normality are being narrowed through emphasis on body optimisation, we need to know what impact this has for the lived experience of those whose bodies deviate from new ideals. Further, what are the micro-practices through which decisions about 'life' and its value are made, prior and subsequent to genetic screening? What are the 'buildings and spaces' involved, and how are they inscribed with 'new' body knowledge? How are notions of 'life' and the imperative of optimisation negotiated in these various sites? And in the case of those faced with the responsibility of literally making life/death decisions – such as the potential parent(s) of a potentially impaired child, following genetic screening – how are the narratives of somatic experts, biological citizenship, faith bodies and the self negotiated? What are the geographies of the commodification and trade in body parts – who and where are the donors, receivers and what is the spatiality of the processes involved? The question of who benefits, who loses needs to be central. Which people in which countries of the world are supporting the search for 'optimisation' in the global north? We have the chance to demonstrate such spatialities, distributive choices, the situatedness of knowledge and the very material and social conditions that mediate the outcomes of advances in technology and scientific medicine. What theory(ies) of power would help unravel differential access to technologies of optimisation and biomedical cure?

The above suggestions are intended to be just that, suggestive rather than a setting out of research directions. They reflect a continued need to broaden what is understood as disability and the processes constructing disability experience, while taking on board an ever shifting context – currently embracing new knowledges and technologies of the body and a climate of social values where an optimal vital capacity is pursued. Working with the body's materiality and its representation in various discursive forms has been an important challenge to disability geography theorising, but has always been located within a sensitivity to the specificities of environments within which materialities and representations are negotiated and experienced. Geographers have much to contribute in elaborating the social and spatial processes that locate 'bodies in context.' Although a normalising discourse of the ideal body, prevalent in the global North but increasingly being adopted across the globe, may be circumscribing a narrower 'normality', it is within the everyday social interactions in a variety of sites – homes, neighbourhoods and cities – where notions of normativity are grounded and given meaning. Normalising discourses also vie with human rights legislation, which advocates for body/mind differences. The embodiment of new knowledge of the body is realised through social and spatial processes at various scales.[3] Scientific knowledge, for instance, is rendered into practical knowledge through codifications and representations informing policy level analysis, which, in the case of the technological intervention of genetic screening, for example, are put into practice through ethical procedures and narratives surrounding testing.[4] The routine practices and interpretation of scientific knowledge occurring in such sites – e.g., government offices and prenatal clinics – are illustrative of the social processes, and their specific spatialities, involved in the reach of biomedicine and its technologies into the re-constitution of understandings of disability. A radical body politics begins here, in analysis of the everyday, routine practices and their effects. The body is a critical site in the normalising practices of medical and scientific advancement and as categorisations shift with changes in knowledge and its codification we would do well to observe how the micropolitics of power play out in those sites where knowledge is made, legitimised and acted upon. And what of the lived experience of the effects of new body knowledges? Disability is a significant dimension of embodied experience, yet individuals occupy multiple social locations and it is important to avoid essentialising identities. In addition to attending to diversity within disability, such as gender and 'race', research needs to consider interdependencies and shared worlds as shifts in the context and meaning of 'life'.

Research in disability geography shows a commitment to social justice issues and a desire for a politics and ethics of research that respects and is inclusive of those who may be affected by research findings. A commitment to research methods that engage users of technology or those dealing with the complex ethical issues involved in attempts to provide a better life for himself or herself or a loved one remains an imperative. Research collaborations with the 'somatic experts' on the front line of working with medical technologies would be useful in bringing a polyvocality to the epistemological, practical and ethical issues involved in clinical education, counselling and reception of genetic and other intervention-related knowledge. There is considerable complexity in the politics and ethics of all disability research, including that of the position of the personal within them (Worth 2008), but issues of power remain central in analysis – particularly in its playing out in the management of 'life itself'. A neglect in research has been that of power relations in the context of global interconnections. One dimension is the impact of global movements of people, as well as of knowledge and technological expertise. Some work has considered the intersectionality of gender, 'race' and

---

3 Scale is a complex concept that has occasioned considerable debate (see Herod 2009). Here I am referring to scale as a 'representational trope' in thinking about ways in which social and processes are linked across space.

4 For example, Browner and Press (1995) describe how state policy on the incorporation of genetic screening into prenatal care is translated into practice through the routine interactions between pregnant women and those administering the diagnostic test. The interactions are shown to vary in terms of information given according to everyday pressures such as time.

disability (e.g., Dossa 2009), but we know relatively little of inequities emerging from the socio-spatial processes of new body knowledges and how these are embodied. A rich research agenda is on hand.

Geographical work has made a considerable contribution to disability research through attention to the spatiality of a range of processes through which disability is experienced and discourses about (dis)abled bodies are embodied. The chapters in this volume represent an array of approaches and topics that reflect the broadening of work on geographies of disability and a continued commitment to a sense that better understanding of marginalising processes can result in improved lives of disabled people and those living with chronic illness. New knowledges of the body open up further the complexity of the interweaving of discourse and materiality in relations of power and the politics of knowledge of the body, located at the interface of biology and culture. New senses of biological citizenship and 'life itself' come with new social justice issues, global in reach. A geographical imagination can contribute much to understanding the new spatialities and sites of regulation of 'optimisation'. But while it remains critical that the message of social construction prevails, it is also important that the biomedical model is retained in analyses. The notion of embodiment, which can integrate the biology of bodies and narrative constructions of 'life' and (dis)ability (which include advances in biomedical knowledge and technologies) through lived experience, remains a useful concept. This acknowledges 'context' in relation to shifting ideas and geographies of dominant narratives, as well as the 'local' social and material environments of people's everyday lives.

References

Agamben, G. 1998. *Homer Sacer: Sovereign Power and Bare Life*. Stanford: Stanford University Press.

Brown, T. and Duncan, C. 2002. Placing geographies of public health. *Area*, 33(4), 361-369.

Browner, C.H. and Press, N.A. 1995. The normalization of prenatal diagnostic screening, in *Conceiving the New World Order: The Global Politics of Reproduction*, edited by F.D. Ginsburg and R. Rapp. Berkeley, Los Angeles and London: University of California Press, 307-322.

Buchanon, A., Brock, D.W., Daniels, N. and Wickler, D. 2000. *From Chance to Choice: Genetics and Justice*. Cambridge: Cambridge University Press.

Butler, R. and Parr, H. (eds) 1999. *Mind and Body Spaces: Geographies of Illness, Impairment and Disability*. London: Routledge.

Chinai, R. and Goswami, R. 2007. Medical visas mark growth of Indian medical tourism. *Bulletin of the World Health Organization* [Online], 85(3), 161-224. Available at: www.who.int/bulletin/volumes/85/3/07-010307/en/index.html [accessed 8 June 2009].

Chouinard, V. 1997. Making space for disabling difference: challenges to ableist geographies. *Environment and Planning D: Society and Space*, 15, 379-387.

Cohen, L. 2003. Where it Hurts: Indian Material for an Ethics of Organ Transplantation. *Zygon: Journal of Religion and Science*, 38(3), 663-688.

Crooks, V. and Chouinard, V. 2006. An embodied geography of disablement: chronically ill women's struggles for enabling places in spaces of health care and daily life. *Health and Place*, 12, 345-352

Dorn, M. and Laws, G. 1994. Social theory, body politics, and medical geography: extending Kearns' invitation. *Professional Geographer*, 46(1), 106-10.

Dossa, P. 2009. *Racialized Bodies, Disabling Worlds: Storied Lives of Immigrant Muslim Women*. Toronto: University of Toronto Press.

Environment and Planning D. 1997. Theme issue on Geographies of Disability. *Environment and Planning D: Society and Space*, 15, 379-480.

Foucault, M. 1973. *The Birth of the Clinic: An Archaeology of Medical Perception*. New York: Vintage Press.

Gleeson, B. 1999. *Geographies of Disability*. London: Routledge.

Hall, E. 2000. 'Blood, brains and bones': taking the body seriously in the geography of health and impairment. *Area*, 32(1), 21-30.

Hansen, N. and Philo, C. 2007. The normality of doing things differently: bodies, spaces and disability geography. *Tijdschrift voor Economische en Sociale Geografie*, 98(4), 493-506.

Herod, A. 2009. Scale: the local and the global, in *Key Concepts in Geography*. 2nd Edition. edited by N.J. Clifford, S.L. Holloway, S.P. Rice and G. Valentine. Thousand Oaks, CA: Sage Publications, 217-235.

Imrie, R. (ed.) 2001. Barriered and bounded places and the spatialities of disability. *Urban Studies*, 38(2), 231-237.

Imrie, R. and Edwards, C. 2007. The geographies of disability: reflections on the development of a sub-discipline. *Geography Compass*, 1(3), 623-640.

Kaw, E. 1993. Medicalization of Racial Features: Asian American Women and Cosmetic Surgery. *Medical Anthropology Quarterly New Series*, 7(1), 74-89.

Kitchin, R. 2000. *Disability, Space and Society*. Sheffield: Geographical Association.

Morris, J. (ed.) 1996. *Encounters With Strangers: Feminism and Disability*. London: Women's Press.

Moss, P. and Dyck, I. 2002. *Women, Body, Illness: Space and Identity in the Everyday Lives of Women with Chronic Illness*. Lanham, MA: Rowman and Littlefield.

Nullis-Kapp, C. 2004. Organ trafficking and transplantation pose new challenges. *Bulletin of World Health Organization*, 82(9), 179.

Oliver, M. 1990. *The Politics of Disablement*. London: Macmillan.

Parr, H. 2002. Medical geography: diagnosing the body in medical and health geography, 1999-2000. *Progress in Human Geography*, 26(2), 240-251.

Petersen, A. 2007. *The Body in Question: A Socio-cultural Approach*. London: Routledge.

Rose, N. 2007. *The Politics of Life Itself.* Princeton: Princeton University Press.

Scheper-Hughes, N. 2000. The Global Traffic in Human Organs. *Current Anthropology*, 41(2), 191-224.

Shakespeare, T. 2005. Disability, genetics and global justice. *Social Policy and Society*, 4(1), 87-95.

Stepnisky, J. 2007. The biomedical self: hermeneutic considerations. *Social Theory and Health*, 5(3), 187-207.

The Times. 2008a. Stem-cell baby born to cure brother of a killer illness sparks clash with Catholics. *The Times*, 18 October, 47.

The Times. 2008b. Body and soul: fresh thinking on disability. *The Times*, 29 November, 5.

Worth, N. 2008. The significance of the personal within disability geography. *Area*, 40(3), 306-314.

Index